Liquid Assets

How Demographic Changes and Water Management Policies Affect Freshwater Resources

Jill Boberg

Prepared for the Compton Foundation

The research described in this report was supported by the Compton Foundation.

Library of Congress Cataloging-in-Publication Data

Boberg, Jill.
Liquid assets : how demographic changes and water management policies affect freshwater resources / Jill Boberg.
p. cm.
"MG-358."
Includes bibliographical references.
ISBN 0-8330-3807-9 (pbk. : alk. paper)
1. Water-supply—Developing countries. 2. Water use—Developing countries. 3. Water resources development—Developing countries. I. Title.

HD1702.B63 2005
363.6'1'091724—dc22

2005010743

Published 2005 by the RAND Corporation
1776 Main Street, P.O. Box 2138, Santa Monica, CA 90407-2138
1200 South Hayes Street, Arlington, VA 22202-5050
201 North Craig Street, Suite 202, Pittsburgh, PA 15213-1516
RAND URL: http://www.rand.org/
To order RAND documents or to obtain additional information, contact
Distribution Services: Telephone: (310) 451-7002;
Fax: (310) 451-6915; Email: order@rand.org

Preface

Human beings have a powerful effect on the environment, as is becoming increasingly clear. Demographic factors are commonly recognized as one of the primary global drivers of human-induced environmental change, along with biophysical, economic, sociopolitical, technological, and cultural factors. Concerns about demographic effects on the environment are fueled by demographic trends such as global population growth and the exponential growth of urban areas. These trends have spawned a body of literature regarding the connections between demographic trends and natural resources, such as water, much of which has taken an alarmist view. The reports often limit themselves to looking at the effects of population growth, and treat water supplies as static and population as ever increasing, inexorably leading to a crisis of water availability. This monograph attempts to present a more holistic view of the interaction between demographic factors and water resources by considering a wider range of demographic variables as well as a set of mitigating factors that influence the availability of water at the local level. The monograph focuses primarily on conditions in developing countries, since that is where the forces of demographics and natural resources intersect with the fewest social and economic resources to mediate their impacts.

This monograph should be of interest to anyone concerned with the interaction between demographic issues and water and other environmental issues, including policymakers and academics.

The funding for this project was provided by a grant to the RAND Corporation from the Compton Foundation. This research was conducted for Population Matters, a RAND project to synthesize and communicate the policy-relevant results of demographic research. Through publications and outreach activities, the project aims to raise awareness and highlight the importance of population policy issues and to provide a scientific basis for public debate over population policy questions.

The *Population Matters* project is being conducted within RAND Labor and Population, a program of the RAND Corporation.

For further information on the *Population Matters* project, contact

Julie DaVanzo
Population Matters
RAND
P.O. Box 2138
1776 Main Street
Santa Monica, CA 90407-2138
Email: Julie_DaVanzo@rand.org

or visit our homepage at http://www.rand.org/popmatters.

Contents

Figures

Tables

Summary

Demographic factors play an important role in environmental change, along with biophysical, economic, sociopolitical, technological, and cultural factors, all of which are interrelated. Recent demographic trends have sparked concern about the impact of the human population on a critical element of the natural environment—fresh water. In the last 70 years, the world's population has tripled in size (Bernstein, 2002) while going from overwhelmingly rural to a near balance of urban and rural—a change that affects both how humans use water and the amount they consume.

In the late 1980s, concern over a potential water crisis began to grow. Much of the resulting literature has taken an alarmist view. Numerous reports sensationalized the so-called water crisis without taking into account the local or regional nature of water resources and the relationship between supply and demand. A number of factors are cited to support the position that the earth is headed toward a water crisis. They include the following:

- The human population continues to grow.
- Water withdrawals are outpacing population growth.
- Per-capita water availability is declining.
- Clean, potable water is less available worldwide.

However, calculations of water resources rely on factors that are difficult to measure. It is important to consider the following:

- Water supply and demand are difficult to measure accurately.
- Water management plays an important role in the supply of and demand for water.
- Population forecasts are changeable.

Given these limitations, predictions of water scarcity may be overstated. At the same time, the risk of a water shortage remains. This risk arises not simply from population growth, but from a host of interrelated factors, including other demographic factors.

Figure S.1 illustrates the framework of analysis for this monograph. In this framework, there is not a direct, linear relationship between demographic factors and water availability, but rather there are numerous mitigating factors, such as resource management, human adaptation, and technological fixes, which mitigate the impact of change in demographic factors or water resources. The framework also illustrates the reciprocal nature of the relationship between water and demographic factors.

Figure S.1
Framework of Analysis

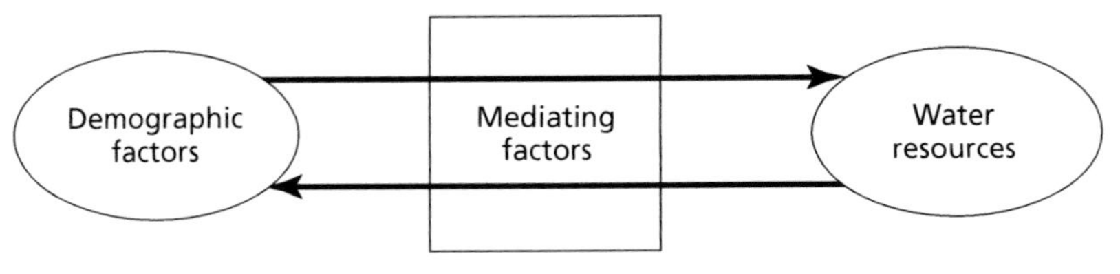

RAND *MG358-S.1*

This monograph integrates literature from different perspectives to explore the factors that cause and mitigate the condition of world water resources and the demographic influences on them, how they interact, and how they work together and against each other.

Freshwater Availability

About 30 percent of the earth's freshwater is found as groundwater, while only about one percent, or about 200,000 cubic kilometers (km^3), of it is easily accessible for human use in lakes, rivers, and shallow aquifers. Humans' primary source of freshwater comes from the water that runs off after precipitation.

Natural Influences on Water Availability

Water is not distributed evenly across land masses, and much of it is far from population centers. About three-quarters of annual rainfall occurs in areas where less than one-third of the world's population lives. Because precipitation is also temporally uneven, people in many regions are unable to make use of the majority of the hydrologically available freshwater supply.

Climate change affects precipitation and therefore affects the freshwater supply. Higher global and regional temperatures mean increased evapotranspiration, changes in snowfall and rainfall patterns, and changes in the intensity, severity, and timing of major storms. These changes will in turn affect the supply of and demand for water, as well as its quality in some regions of the world.

Calculating Water Availability

The quantity of water *practically* available to a population is defined by factors such as quality of water demanded, cost, environmental effects, political and legal agreements, individual wealth, and the technical ability to move water from place to place. It also depends on the quality of water demanded; for example, drinking water must be of higher quality than water for industrial or recreational uses.

The actual amount of water available to any person or group for a particular use depends not only on physical availability, but also on management and infrastructure to capture runoff and groundwater. Simple measures of water availability can hide factors such as variability of access, temporal and spatial availability, and legal and political restrictions, over- or underemphasizing existing or potential signs of stress, and misleading policymakers and water managers.

Patterns of Demand for Freshwater

Water usage is commonly analyzed according to three categories or sectors: agricultural, industrial, and domestic or municipal. Water usage in each sector varies across regions, over time, in relative quantity, and may vary seasonally.

Agriculture is generally the largest user of freshwater, accounting for about 70 percent of all annual water withdrawals worldwide, though in Europe it ranks behind industry. Most agricultural water use is for irrigation. Much of the demand for irrigation is being met in nonsustainable ways (e.g., by pumping nonrenewable groundwater supplies) and therefore may be problematic in the future. Furthermore, much of the future food production in developing countries is expected to come from irrigated land.

Industry is the second largest consumer of water, responsible for about 20 percent of annual worldwide withdrawals. Water is used in industry for cooling, cleaning, processing, generating steam power, and transportation. Because industries tend to cluster in urban areas, industrial water withdrawals are a significant component of urban water demand. There are a growing number of industries in rural areas, however, adding to rural industrial demand.

The domestic and municipal sector generally demands the smallest share of water (about 10 percent worldwide), except in countries with little agriculture or industry. Water is used in the domestic and municipal sector for purposes such as drinking, food preparation, and sanitation. The domestic sector, although less demanding in terms of volume than the other sectors in most places, warrants special attention because of its implications for health and mortality. Safe drinking water is an important public health and political concern. Poor water quality is a major causal factor in infant mortality, premature mortality, and lost productivity in many countries, especially in the developing world.

Demographic Influences on Water Resources

Demographic factors affect water consumption and the quality and health of natural ecosystems both directly and indirectly. Population growth, urbanization, and migration strongly affect the sustainability of water resources. Other factors that affect water use are indirectly linked to population—income levels, a rise in living standards, modifications to landscapes and land use, contamination of water supplies, and inefficiency of water use caused by a failure to manage demand.

Population Size and Growth

Population size is fundamentally linked to water use, and some perceive it to be the most important demographic factor affecting water resources. Although populations and individuals use resources differently, population size and changes in population size affect the magnitude of water use and can affect water quality because humans compromise the quality of water when they increase in number, migrate, and change the landscape.

The 20th century saw world population grow from 1.65 billion to 6 billion people, with almost 80 percent of that increase taking place after 1950. This rapid growth was due in part to mortality reductions, which were partly due to improvements in water quality. Population growth will continue for some time into the future. The water demands of humans will continue to affect water resources in ways we can only partially predict.

Number of Households

The number of households is increasing worldwide, and the average household size is decreasing—both factors that influence water demand. In fact, an increase in smaller households may do more environmental damage than simple population growth. The decrease in household size affects household-related economies of scale and reduces the effectiveness of investments in technical water-saving measures. Per capita, smaller households consume more water and produce more waste. More households require more housing units, increasing the materials needed for construction and contributing to urban sprawl. This damages water resources and water quality by paving over land that would otherwise help filter the water that replenishes lakes and rivers.

Urbanization

Urbanization, both by the degree and the rate of growth, can affect the levels of per capita water use, overtax water resources by concentrating demand in a small area, and overwhelm the existing infrastructure. It can also have a broader impact when freshwater resources must be transported from elsewhere.

Economic Development

A country's level of freshwater use is tied to its level of economic development, and the correlation is strong enough that water use is often used as a measure of economic development. Overall, developing countries use less water per capita than industrialized countries. Freshwater is used to run industry, grow food, generate electricity, and carry pollution and wastes, so increased industrialization is associated with increased water use per capita. Both the individual water use levels and the water used by society to support individuals and their ways of life are higher in industrialized countries.

Within countries, income level also determines the structure of freshwater demand. The poorest people in a country tend to use less water than those with high incomes. In cities, wealthier individuals are more likely to be connected to piped water supplies and waterborne sanitation systems, as well as to consume larger amounts of products that require water to produce. Rural dwellers, who are more likely to be poor, often use less water because of inaccessibility of supply and lower consumption of goods and services.

How Water Resources Influence Demographic Factors

Just as demographic factors affect water resources, water resources also affect demographic variables.

Morbidity and Mortality

Those who have less access to water and sanitation systems—whether due to scarcity or lack of infrastructure—tend also to have higher rates of disease and mortality, both because of poorer drinking water quality and because of reduced availability of water for hand washing. Countries with higher national incomes have built sewer networks and wastewater treatment facilities that have reduced the incidence of waterborne diseases. In contrast, rapidly expanding cities in much of the developing world have not been able to build such infrastructures, and they have higher rates of illness from such diseases.

Migration

Water resources often affect migration decisions. Lack of water can be a push factor in the decision to migrate away from a location. Deforestation, desertification, drought, and lack of land to cultivate can significantly influence migration from rural areas to towns. Millions of people have become refugees due to dam building, environmental deterioration, or the destruction of environmental resources that they require to live or make a living.

Water can also be a pull factor that induces people or industry to move to a region. For example, an industrial entity may choose to locate itself near water resources, and the employment opportunities it creates may influence the scale of future labor migration.

Approaches to Sustainable Water Management

Whether or not a water crisis is imminent, policymakers should explore ways to reduce the pressures on the water supply. The management of water supply and demand can make large differences in water withdrawals, and can influence the quality of water and its impact on human health. Supply management involves the location, development, and exploitation of new sources of water. Demand management involves the reduction of water use through incentives and mechanisms to promote conservation and efficiency.

Supply Management

Options for increasing water supplies include building dams and water-control structures, watershed rehabilitation, interbasin transfers, desalination, water harvesting, water reclamation and reuse, and pollution control.

Dams and water-control structures are used for energy production, flood control, and water storage. Substantial benefits may accrue from large dams, such as electric power production, irrigation, and domestic water provision. However, dams also have large environmental, social, and demographic consequences, including the submergence of forests and wildlife, greenhouse gas emissions from the decay of submerged vegetation, the impact of changed flow in the dammed river, the relocation of families, the destruction of villages and historic and cultural sites, and the risk of catastrophic failure. Currently, there are more than 41,000 large dams in the world, impounding 14 percent of the world's annual runoff.

Reforestation of degraded and unproductive land can be cost effective in the long term if all the ensuing benefits to the water supply, agricultural revenue, and pollution reduction are taken into consideration. However, the high initial cost combined with the drive to find short-term solutions make watershed rehabilitation a less appealing choice in many regions.

Small-scale irrigation systems may provide an opportunity to expand irrigation with fewer constraints than large-scale water projects. Studies have shown that small-scale irrigation systems may be as successful as large-scale irrigation projects, but their success depends on institutional, physical, and technical factors, including technology, infrastructure, and implementation.

Groundwater is a more efficient source of irrigation water than open canals because the water can be accessed when and where it is needed, reducing transportation costs. However, sustainable use of groundwater requires oversight to ensure that the amount tapped does not exceed the amount recharged through the hydrological cycle each year.

Interbasin transfers and water exports are two more methods of providing water to areas with dense populations or insufficient supplies of water. The infrastructure required for interbasin transfers (i.e., using canals or pipelines to transport water), however, is expensive and potentially environmentally harmful, both to the freshwater ecosystem from which the water is extracted and the lands over which the pipes or canals must flow. The selling of water through exports has been controversial because water is perceived as a common resource that should not be sold by private companies.

Water reallocation from one sector to another can be done via supply management, i.e., by reallocating water from the top down, or via demand management, i.e., using incentives to move water between sectors. Because water in developing countries is predominantly consumed by the agriculture sector, reallocation of even small percentages of water consumption to the domestic sector can fulfill domestic needs.

Desalination, a process of turning saltwater into fresh drinking water through the extraction of salts, is an expensive, energy-intensive process, and it produces brine that must be disposed of carefully to avoid environmental damage. For these reasons, it is currently an unappealing option for increasing the water supply, except in the most arid areas of the world.

Water harvesting, i.e., the capture and diversion of rain or floodwater, can provide environmental benefits by reducing the polluted runoff from urban areas into surface waters and can increase efficiency, productivity, and soil fertility on a local and regional basis.

Water reclamation and reuse have helped reduce water use in the industrial sector in some developed countries, and they reduce the amount of pollution released into receiving waters such as oceans, lakes, and streams. However, the practice is limited when the water contains high levels of salt or heavy metals.

Pollution control, i.e., preventing contamination from agricultural residues, soil erosion, urban runoff, industrial effluent, chemicals, excess nutrients, algae, and other pollutants, is another supply option to maintain the quality of the current supply of water. Pollution-control laws in developed countries have helped to clean up rivers, lakes, and streams, and they promote conservation and the efficient use of water. Although most countries have pollution-control laws, many developing countries lack the political will or financial resources to enforce them.

Policy Options for Demand Management

Although supply management measures like those discussed above will help to meet demand for water in the future, much of the effort to meet new demand will have to come through demand management, including conservation and comprehensive water policy reform. Management approaches such as integrated water resources management and demand management offer effective means of providing water for human use while easing the stress on the freshwater ecosystems and the ecological goods and services that they provide.

Poor governance is at the core of many water problems, especially in developing countries. Several categories of policy instruments can be used to enact demand management and improve governance of water systems, including institutional and legal change, market-based incentives, nonmarket instruments, and direct interventions.

Institutional and legal change of the atmosphere in which water is supplied and used includes water-quality matching, decentralizing supplies, and privatization of water management. Water-quality matching involves channeling pure water only to functions that require it, such as human consumption, and using lower-quality water for appropriate uses, such as irrigation and industrial processing. Decentralization allows local communities to control their own water supplies. Privatization involves private firms in the building and operation of water-management systems. Privatization is controversial because private firms are motivated by profits, and concerns about equitable access to water for the poor, the integration of environmental concerns, and the sharing of risks must be addressed before privatization becomes a widely viable option.

Market-based incentives use economic means to modify the behavior of consumers. Water pricing, the most common example of a market-based incentive, typically reduces demand, increases conservation, encourages reallocation between sectors, and increases environmental sustainability by reducing the strain on ecosystems.

Nonmarket instruments, policies that use direct, non–incentive-based laws and regulations to control water use, include quotas, licenses, pollution controls, and restrictions. Educational measures that encourage water conservation are important companions to nonmarket instruments.

Direct intervention, i.e., technological and other interventions to conserve water include, for example, leak detection and repair programs, investment in improved infrastructure, and conservation programs. Conservation can occur through measures such as efficiency standards for plumbing fixtures or changing industrial processes to use less water.

Will There Be a Global Water Crisis?

There will undoubtedly continue to be localized problems of water scarcity, and perhaps more widespread problems in some areas, depending on local physical, social, economic, and cultural conditions. However, a global water crisis can be averted. There are many options for improving water management and alternatives for meeting supply and demand, even for growing and changing populations. Attention to demographic factors is an important part of the formula for staving off water crisis indefinitely. To be successful, there must be sufficient institutional, intellectual, and administrative capacity to create solutions and to execute them.

Sustainable water development and management requires the integration of social and economic concerns with environmental concerns. This effort will be enhanced by research that focuses on as small a scale as possible, to provide data and information on methods that will help to manage water at socially relevant levels, more locally, as well as at environmentally relevant levels such as the watershed. Looking ahead, research on demographic variables that are less understood, such as the impact of the number and size of households on the environment, will help to improve the overall understanding of the relationship between demographics and water resources.

Acknowledgments

The author is indebted to all who helped in the writing of this monograph. Helpful comments on previous drafts and discussions on the crafting of the monograph were gratefully received from formal and informal reviewers Martha Geores, Debra Knopman, Dave Adamson, Mark Bernstein, and Kathleen Micham. Many thanks to research communicator Jennifer Li, along with Tania Gutsche, Megan Beckett, and Rebecca Kilburn.

For her extreme patience and always insightful guidance, copious notes, and detailed suggestions, special thanks and appreciation go to Julie DaVanzo.

CHAPTER ONE
Introduction

The role of demographic change—that is, change in population size, composition, and distribution—in environmental change is widely acknowledged to be an important one, but the specific relationship between population and the environment is still incompletely understood. Demographic factors as a whole are commonly recognized as one of the primary global drivers of human-induced environmental change, along with biophysical, economic, sociopolitical, technological, and cultural factors (Orians and Skumanich, 1995; WWC, 2000; Hunter, 2000). But these factors do not exist in isolation—the influence of each of these factors on the environment is interrelated with the others. For example, an increase in population in a region may affect the environment differently depending on the economic resources of the people and governments in the region.

Concerns about demographic effects on the environment are fueled by recent demographic trends. For example, in the last 70 years, the world's population has tripled (Bernstein, 2002). At the same time, the population of the world has gone from overwhelmingly rural to a near balance of urban and rural (UN, 2001). Because the use of demographic data by environmental policymakers is sometimes spotty or inconsistent, it is worthwhile for demographers and those setting the environmental policy agenda to share their understanding of the connections between demographic factors and the environment (Orians and Skumanich, 1995). In this effort, it is important that both the direct and indirect implications of demographic factors be studied to understand the multifaceted relationships between population and an environmental resource such as water.

Water is a basic need of both humans and the ecosystems on which they depend. It is central to life requirements, as well as to agriculture, energy, and industrial production. It is equally central to the well-being of the natural world—the water cycle is the engine for life on earth. Because water quality and quantity are crucial to life, policymakers need to understand the reciprocal relationship between population change and this important resource.

"Humans have always discovered, diverted, accumulated, regulated, hoarded, and misused water" (de Villiers, 2000, p. 46). From early in human history, water engineers (or their equivalents) have harnessed supplies of water to meet the ever-increasing demand predicted by planners. In recent times, water works, some of unprecedented size (e.g., Hoover Dam in the 1930s, the California Aqueduct in the 1950s, Aswan Dam in the 1970s), were built, and water flowed. It seemed that this model, one of priority of water supply over concerns such as the environment or reduced water use, would continue to prevail indefinitely. In the late 1980s, however, sustained widespread concern about water and the potential for crisis began gaining currency. Shiklomanov's 1990 research detailing his calculations on global water resources sparked a number of papers using those figures and recent population forecasts to draw conclusions about the future of water supplies and a looming water crisis. In 1992, hydrologist Malin Falkenmark pioneered the concept of a water stress index and thresholds of water stress and scarcity (Falkenmark and Widstrand, 1992). Many papers followed refining these concepts and popularizing the notion of water scarcity (e.g., Postel, 1997; Serageldin, 1995; Abramovitz and Peterson, 1996; Gleick, 1993). As a result, water availability has become a more widely discussed issue for many policymakers and academics in fields, such as demographics, that may not have previously considered it.

Unfortunately, much of this literature has taken an alarmist view. Numerous reports by academics and advocacy groups have sensationalized the so-called water crisis without, perhaps, a complete understanding of the characteristics of water use and availability, such as its often local or regional nature and the relationship between demand for and supply of water resources. They often treat water supplies as static and population as ever increasing, inexorably leading to a crisis of water availability. Although it is true that population growth puts pressure on water and other natural resources, it is not a simple cause-and-effect relationship. Many mitigating factors are not given the attention they deserve. The emphasis on water crisis in a number of recent papers is explained partly by the fact that, on the whole, research that links population and environmental change, including water resources, limits itself to looking at the effects of population growth (Pebley, 1998). It is important that other demographic variables such as urbanization rates, income levels, and numbers of households, for example, as well as natural resource consumption, be taken into account, and that the reciprocal impacts of environmental degradation on reproduction, fertility, migration, and mortality are considered (Pebley 1998). Additionally, much of the attention concerning the relationship between population and water has centered on domestic water use. This is despite the fact that in most of the world, domestic water use accounts for only a small portion of total water use. The agricultural and industrial sectors use far more water than the domestic sector but are impacted differently by demographic factors, including population growth.

The following list catalogs the factors linked in many reports to justify heightened attention to water availability. This is followed in the next section by a list of ways in which our knowledge for making such judgments is incomplete. As suggested, and as will be discussed later in the monograph, not all of these factors should necessarily be a cause for apprehension—some are based on unreliable statistics, others on data so aggregated as to be useless in an analysis of water resources, and so on. However, they are all used to encourage a crisis mentality for policy makers when considering water availability in the future.

These factors are as follows:

- **The human population continues to grow.** Population growth and human activities inevitably affect both the quantity and quality of resources. Population growth is often seen as the engine for resource use. Human populations use water resources in many ways, including as repositories for the effluents generated by production and consumption processes. Because of this, and because of the rate and spatial dimensions of population growth, there has been an unprecedented increase in the impact of human populations on water resources. There is concern, expressed in many published reports, that there will not be enough water to maintain the larger human populations predicted for the future. A related matter is whether the growth and distribution of the population will destroy the freshwater ecosystems that maintain life on earth.

- **Water withdrawals are outpacing population growth.** The tripling of the human population in the past 70 years has been accompanied by a six-fold increase in water withdrawals (Gleick, 1993). It is estimated that withdrawals of freshwater for human use have risen almost forty-fold in the past 300 years, and that over half of that increase has come since 1950 (Abramovitz and Peterson, 1996). Although the validity of the information may be in question, many reports have also expressed concern over an often-reported statistic of uncertain origin stating that humans currently use 54 percent of all accessible surface water runoff from rivers, lakes, streams, and shallow aquifers—about twice as much as we used 35 years ago. This statistic is accompanied by the prediction that by 2025, population growth alone may cause an increase in human water appropriation to 70 percent of accessible surface water runoff (see, for example, Cosgrove and Rijsberman, 2000; Hinrichsen, Krchnak, and Mogelgaard, 2002). Additionally, more and more nonrenewable groundwater supplies are being tapped and depleted or polluted before they can be used. In Saudi Arabia, for example, at least half of the country's known fossil (nonrenewable) water reserves have been consumed in the years since 1984 (Smith, 2003). It is estimated that 36–48 cubic miles more groundwater is pumped out of aquifers than is recharged naturally, resulting in water table declines of 33–164 feet in cities worldwide in the years up to

1998 (Hinrichsen, Krchnak, and Mogelgaard, 2002; Foster, Lawrence, and Morris, 1998). Although these numbers do point to some alarming trends, the implication that increased population leads inevitably to increased demand for water is false, as will be discussed later.

- **Per-capita water availability is declining.** In per-capita terms, water availability is declining rapidly, from 17,000 cubic meters (m^3) per person in 1950 to 7,044 cubic meters in 2000 (WRI, 2000). Currently, by some definitions, water stress and scarcity exist in many places in the world, and many experts predict widespread water scarcity over the next century (Revenga, Brunner, et al., 2000; Gardner-Outlaw and Engelman, 1997; UN CSD 2001). Hydrologists often describe water stress as water supplies of between 1,000 and 1,700 m3 per person per year. This is thought to represent a situation in which disruptive water shortages may occur frequently. A country is said by some experts to be in water crisis if it has supplies of less than 1,000 m3 per person per year, a situation that is predicted to lead to problems with food production and economic production. Table 1.1 shows those countries that were in water stress or crisis in 1995 as defined in this way. The population in these countries represents about 11 percent of the world's population. Despite these alarming numbers, however, the level of aggregation hides an enormous amount of variability. As is discussed later in this report, the measures described here are too simplistic to be useful in any meaningful way.

Table 1.1
Countries with Estimated Annual Per-Capita Water Availability of 1,700 m^3 or Less in 1995

Country	Per-Capita Water Availability (m^3)
Middle East and North Africa	
Qatar	91
Libya	111
Bahrain	162
Saudi Arabia	249

Table 1.1—continued

Country	Per-Capita Water Availability (m^3)
Jordan	318
Yemen	346
Israel	389
Tunisia	434
Algeria	527
Cyprus	745
Oman	874
United Arab Emirates	902
Egypt	936
Morocco	1,131
Kuwait	1,691
East Asia and Pacific	
Singapore	180
South Korea	1,472
Latin America and the Caribbean	
Barbados	192
Haiti	1,544
Peru	1,700
Europe and Central Asia	
Malta	367
United Kingdom	1,222
Poland	1,458
Sub-Saharan Africa	
Burundi	594
Cape Verde	777
Kenya	1,112
South Africa	1,206
Rwanda	1,215
Somalia	1,422
Comoros	1,667

SOURCE: Gardner-Outlaw and Engelman (1997).

- **Clean, potable water is less available worldwide.** Per-capita availability of clean, potable water is declining nearly everywhere in the world, even where water scarcity is not a problem. The

> problem with availability arises not only due directly to numbers of people but also indirectly due to the eroded silt, sewage, industrial pollution, chemicals, excess nutrients, and algae that degrade much of the available water. Human populations are affected by poor water quality, resulting in increased morbidity and mortality. Poor water supply, sanitation, and personal and domestic hygiene were held responsible for 5.3 percent of global deaths and 9.4 percent of all premature deaths (those occurring before they would be statistically expected) in 1990 (Murray and Lopez, 1996). The mortality impact of poor water quality and sanitation were not distributed evenly, as diarrheal diseases, transmitted primarily through fecal contamination of water and food, were responsible for 2.2 million deaths in low-income countries, compared to about 7,000 deaths in high-income countries in 1998 (WHO and UNICEF Joint Monitoring Programme for Water Supply and Sanitation, 2000). The heaviest burden of mortality is borne by children under five, especially in developing countries.

In competition for freshwater, wildlife and freshwater ecosystems are usually the losers, and when aquatic ecosystems are undermined, human prospects are equally threatened (Hinrichsen, 2003). The problem with supply becomes a broader issue if the supply of water to nonhuman users is taken into consideration. Although less is known about the exact water quality and quantity requirements of freshwater ecosystems and the flora and fauna that form them, there is a critical need for clean, plentiful, and free-flowing water for wildlife (Hinrichsen, Krchnak, and Mogelgaard, 2002). Currently, human modifications of the landscape affect the hydrological cycle, draining wetlands, and disconnecting rivers and streams from their floodplains. This fragmentation affects 60 percent of the world's largest 227 rivers, and has serious effects on freshwater ecosystems (WCD, 2000). For example, more than 20 percent of freshwater fish species are currently on the threatened or endangered list, or recently extinct (Ricciardi and Rasmussen, 1999).

On the surface, the problem described is a classic one of supply and demand. Freshwater supply is finite and shrinking, due to pollution and non-renewable groundwater mining. Demand, on the other hand, is currently growing, as population, development, urbanization, and incomes increase. With the accompanying changes in distribution and income, population influences both supply and demand. This underlines the importance of population in the matrix of resource management, but does not reflect the complexity of either the demographics or the resource and its management.

Fallacies in the Prediction of the Water Crisis

Although there is a vast body of research on the sustainability of water resources, certain variables make it difficult to predict the future of water resources conclusively. Because calculations of water resources rely on factors that are difficult to predict or measure, those who wish to predict the future of water at either the global or more local level face a number of challenges. These challenges point out some of the fallacies in the predictions of water crisis that have been popular in recent years.

Water Scarcity Cannot Be Defined by a Single Number

Defining water scarcity as an upper limit for water availability per capita seems straightforward, but scarcity is more than simply a shortage of water in time and space (Turton and Warner, 2002). Water scarcity, instead, may be defined as a condition in which demand for water exceeds the prevailing level of local supply (Turton, 1999). Because demographic variables strongly influence demand for water, it can be argued that water scarcity is, largely, economically and socially determined. That is, beyond the three liters or so a day required for survival, demand and need depend on culture-based habits of water consumption. Additionally, low-quality water can be used for fewer functions than can high-quality water. Thus, the degradation of water quality can be considered another cause of water scarcity (Turton and Warner, 2002). Even if freshwater is available, if it is unsafe to use, scarcity can occur. This is important, because it points out the error of a single number being used, as it often is, to define water scarcity.

Scarcity also depends on a society's ability to use resources, suggesting three types of water scarcity: absolute (limited by technology), economic (limited by economic choices), and induced (limited by political choices or ineptitude) (Sexton, 1992). Genuine or absolute water scarcity is a function of the hydrological climate, and water users must adapt to it. In economic water scarcity, there can be plenty of water on average in a country, but it may be located far from population centers and other areas where it is needed (Gardner-Outlaw and Engelman, 1997). Human-induced water scarcity is exacerbated by human behavior and can be minimized and controlled (Falkenmark, 1994). This typology suggests that simple demand-supply equations based on generalized scarcity measures will inaccurately portray the local nature of a particular water situation.

Water Supply and Demand Are Often Difficult to Measure Accurately

Data on water availability are often limited and usually reported on an aggregated scale. In more developed countries, precipitation, temperature, and runoff are generally well measured. However, in other regions of the world, similar data are difficult to obtain, and some governments are loath to share the data they have. Additionally, watershed data, which are the logical basis for measuring water supply, are hard to come by, as data collection tends to follow political, rather than ecological, borders. Data on water demand are even less available than data on supply. Household use is usually estimated rather than measured precisely, especially in places where water is collected, not purchased, and the actual uses for the water are generally unknown. Even industrial use is rarely inventoried, and data on agricultural water use are uneven and unreliable (Gleick, 2002). This is also true for rainfall and groundwater, which are less likely to be traded as commodities. Many tools are used for estimation of these supply and demand data, but in the absence of good primary data, as is the case in many developing countries, even these estimates can be seriously flawed.

Water Management Is Critical

In the same vein, the supply and demand for water, and therefore its abundance or scarcity, depend significantly on the management of the resource and its use. For example, the timing of water availability is critical in many cases. Much of the world has distinct rainy seasons, where there is an abundance of rain followed by an almost complete absence of precipitation. In these places, the timing and volume of the rains can make or break a farming season—the same amount of water can be either a benefit or a detriment depending on timing. Drought may be a part of the normal climatic cycle, and ecosystems and humans have adapted to this variability. The management of water supplies in these situations is critical—poor management may create functional water scarcity in a country with seemingly abundant supplies of fresh water.

Population Forecasts Are Changeable

The United Nations made predictions of total global population in the year 2000 at eight different times from 1973 to 1998. The predicted 2000 populations ranged from about 6.25 billion (predicted in 1990) to about 6.05 billion (predicted in 1998). The actual population in 2000 was just over 6.05 billion. In the same period, similar predictions for the 2000 population in less-developed countries varied from about 4.85 billion to just under five billion. The actual 2000 population of less-developed countries was about 4.87 billion. In general, projections overstated global population numbers by as much as 200 million people, over three percent. These projection errors were a result of the underestimation of declines in fertility rates and the death toll due to AIDS, especially in developing countries. Predictions of subnational population and other demographic changes are even more difficult, primarily due to lack of data on local demographic trends. Since the most appropriate level for the analysis of water supply and use is subnational, this lack of data and potential for error can adversely affect resource planning. Prediction errors may be further amplified by the use of nontraditional measurement units such as watersheds, arguably the most appropriate unit of measurement for water resources. Prediction errors, while inevitable, can make very significant differences in projections of other trends that are based on them. Researchers try to deal with these errors by giving a range of predictions, but there is always a question of which numbers to use in subsequent calculations, and of compounding the errors in subsequent error ranges. If projections of water supply and use, and therefore predictions of water crisis, are based on these aggregate numbers, as they often are in papers on the global water crisis, the changes may impact projections for water availability, as well as food requirements and other indirect water uses.

The Future of Water Resources

Given these limitations, like earlier predictions for population, energy, and other global problems, predictions of water scarcity may be overstated. Forecasts are made using historical data and assumptions about what the future will look like. The most likely future looks like a continuation of the past, so the most obvious presumption is that resource use that has increased at a steady rate in the past will continue to do so in the future. Scholars from Malthus to the Group of Rome have predicted increased population and scarcity that would devastate nature and society. In each case, the improvement of technology, the fluctuations in demand due to economic and social conditions, the change in population growth rates, and especially the vagaries of human adaptation to actual or perceived scarcity have confounded the predictors.

However, water is different from other threatened resources in a number of ways. First, it has no real substitutes, and we cannot create more of it. We can make freshwater out of seawater, but only at relatively high economic costs. Second, it is a localized resource, one that is relatively difficult to transport long distances. Although technology has solved the transport problem to some extent, there are severe economic and environmental constraints to such transport. One exception to this is that water can be indirectly exported in the form of food grown through its use. Third, freshwater is an integral part of the ecosystem from which it is drawn, and the destruction of ecosystems by the depletion or sullying of freshwater sources affects all life on the planet.

There are many places in the world where access to water is difficult, where water is being used wastefully, or where freshwater ecosystems are being destroyed. Most of these water-challenged areas are local, but some of these water flash points are more widespread. They include parts of Africa, the Middle East, and China. According to de Villiers (2000, p. 23), the criteria indicating that there may be a current or looming water problem include the following:

- The accessible water supply is static or falling.
- Substantial water supplies are acquired from outside a country.
- The area is arid or has unreliable rainfall with a low storage capacity.
- The population is increasing.
- There are competing, incompatible demands for water (agriculture, basic domestic needs, industry).

These additional factors were left off this list:

- the use and depletion of nonrechargeable or fossil groundwater
- the influence of resource management practices, pollution, watershed management, and ecosystem health on water resources
- the influence of demographic factors besides population growth.

These additional factors can create a water crisis where there would otherwise be none, or avert a crisis that would exist under another scenario. The existence of all or some of these criteria does not necessarily mean that a water crisis is inevitable, but does indicate a risk for problems. In any case, a water crisis appears to be the result of a number of interrelated factors, including demographic factors, and not a direct result of population growth.

A simple framework arises from this observation. Illustrated in Figure 1.1, this framework links water resources and demographic factors, and points out the influence of mitigating factors on the relationship. The framework emphasizes the view that there is no direct, linear relationship between demographic factors and water resources and availability, but that there are one or more mitigating factors such as resource management, human adaptation, technological fixes, and others, which mitigate the impact of changes in demographic factors or water resources. It also illustrates the reciprocal nature of the relationship between water and demographic factors, highlighting the existence of mitigated cause and effect that can emanate from either source—not only the effect of changing demographic factors on water resources, but also the impact of water resource dynamics on demographic factors.

The monograph explores the factors that cause and mitigate the status of world water resources and the demographic influences on them, how they interact, and how they work together and against each other. First, it reviews the amount of freshwater available on earth and the availability of that water for human use. Then it discusses patterns of human demand for fresh water. Next, it explores key demographic factors such as population size, composition, and mortality, and how these factors influence and are influenced by freshwater resources. Following that, it describes water management options that may help sustain the earth's supply of freshwater. Finally, it presents conclusions on the sustainability of water resources and directions for future research. In doing so, this monograph endeavors to integrate literature originating from different perspectives—population-centered, water-centered, global, and local.

Although worldwide water and population factors are considered, the monograph focuses more closely on the problems issuing from such interactions in developing and other lower-income countries. This focus stems from the disproportionate extent of problems experienced by low-income countries associated with both demographic and resource management concerns. As will become clear, the greatest amount of population growth, urbanization, and other

Figure 1.1
Framework of Analysis

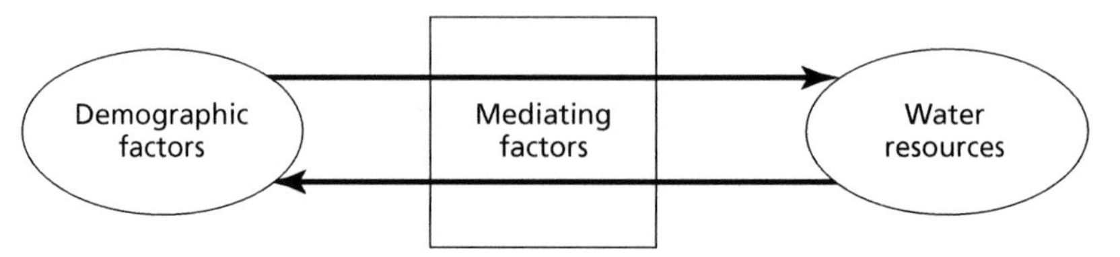

RAND *MG358-1.1*

demographic changes are taking place in lower-income countries. In many cases, too, lower-income countries are located wholly or partly in places that are not well endowed with water and other natural resources. These countries also in many cases have fewer social and economic resources to apply to the solution or avoidance of problems associated with demographic-resource interactions.

CHAPTER TWO

Freshwater Availability

Earth is called the water planet, and 71 percent of its surface is covered with water. In addition to the water on the surface, there is water in underground aquifers, glaciers, icecaps, and the atmosphere. But very little of this water is available for human use, even including that amount needed for ecosystems to function properly. This chapter describes the earth's water cycle, freshwater ecosystems, and the processes that naturally recycle freshwater on the earth. It also discusses water availability and the natural and human causes for changes in the water supply.

Quantity of Water on Earth

Although the total amount of water on earth is massive, estimated at some 1.4 billion cubic kilometers (km^3), the majority (97.5 percent) of it is saltwater (Figure 2.1). Of the 2.5 percent that is freshwater, an estimated stock of approximately 35 million km^3, most of it is tied up in ice and permanent snow cover. Overall, about 30 percent of the earth's freshwater is found as groundwater, while only about one percent, or about 200,000 km^3, of it is easily accessible for human use in lakes, rivers, and shallow aquifers (Shiklomanov, 1993, 2000). Humans' primary source of freshwater comes from the water that runs off after precipitation. This runoff plus groundwater recharge equals about 40,000 to 47,000 km^3 per year (Gleick, 2000b). However, the

Figure 2.1
Earth's Supply of Water

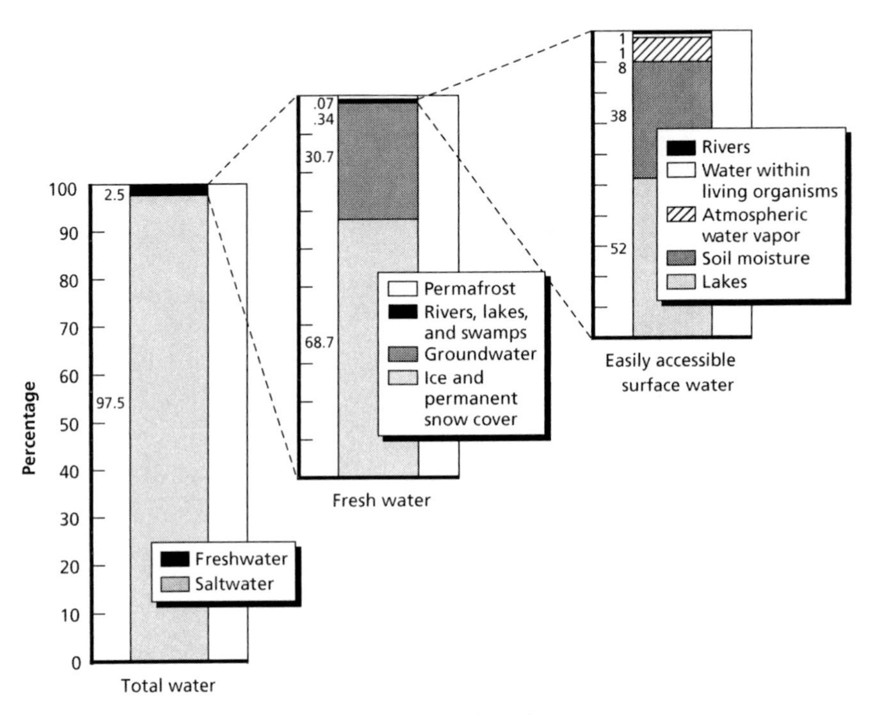

SOURCE: Hinrichsen, Krchnak, and Mogelgaard (2002).
RAND *MG358-2.1*

use of fossil groundwater and other water sources that are relatively slow to be replenished is common and growing.

The Hydrological Cycle and Sustainable Water Use

The hydrological cycle is the repeated process of the evaporation and redistribution of water in various forms around the earth. The complete cycle and the approximate amounts of water involved in each stage are shown in Figure 2.2. The annual precipitation on earth is more than 30 times the atmosphere's total capacity to hold water, which means that water is recycled relatively rapidly between the earth's surface and the atmosphere.

The energy of the sun evaporates water into the atmosphere from oceans and land surfaces. Evaporation is the change of liquid water to a vapor. Sunlight aids this process as it raises the temperature of liquid water in oceans and lakes. As the liquid heats, molecules are released and change into a gas. Warm air rises up into the atmosphere and becomes the vapor available for condensation. Some of the earth's moisture transport is visible as clouds, which themselves consist of ice crystals or tiny water droplets. The jet stream, surface-based circulations like land and sea breezes, or other mechanisms propel clouds and vapor from one place to another until they are condensed back to a liquid phase. Water then returns to the surface of the earth in the form of either liquid (rain) or solid (e.g., snow, sleet) precipitation. Some water on the ground, in streams, or in lakes, returns to the atmosphere as vapor through evaporation, and water used by plants may return to the atmosphere as vapor through transpiration, which occurs when water passes through the leaves of plants. Collectively known as evapotranspiration, both evaporation and transpiration occur at their highest rates during periods of high temperatures, wind, dry air, and sunshine.

Although the total amount of water on earth is fixed, the physical state of the water, on a time scale of seconds to thousands of years, is continuously changing between the three phases (ice, liquid, and water vapor), circulating through the different environmental compartments (ocean, atmosphere, glaciers, rivers, lakes, soil moisture, and groundwater), and renewing the resources. Average replenishment rates show considerable range, from thousands of years for the ocean and polar ice down to biweekly or even daily replenishment of water in rivers and in the atmosphere (EEA, 1995).

Average annual global evaporation from the ocean is six times higher than evaporation from land, while precipitation over the ocean is 3.5 times higher than over land. Based on current estimates, this results worldwide in approximately 40,000 to 45,000 km^3 of water

Figure 2.2
The Hydrological Cycle

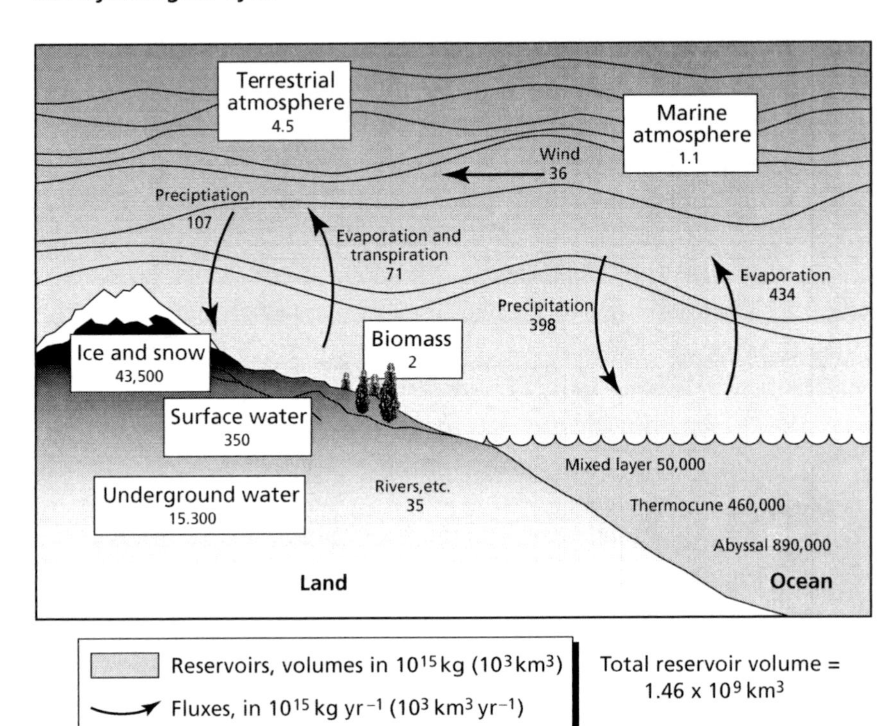

SOURCE: NRC.
RAND *MG358-2.2*

each year transported from the ocean via the atmosphere to renew the freshwater resources (EEA, 1995). The balance of water that remains on the earth's surface is runoff, which empties into lakes, rivers, and streams and is carried back to the oceans, as well as recharging groundwater, and is potentially available for consumption each year (Institute of Water Research, 1997). This is the renewable supply of water, that part of the hydrological cycle that can be utilized each year without leading to the depletion of freshwater resources; it does not include groundwater resources that are no longer replenished.

This renewable resource is referred to as "blue water," while "green water" refers to the rainfall that is stored in the soil and evaporates from it. Green water is the primary source of water for rain-fed agriculture and for freshwater ecosystems, and is based on the level and flow of water in soil, which depends on soil texture and structure as well as climatic factors. It is significantly affected by land use and changes in land use, which are in turn affected by demographics (FAO, 1999). About 60 percent of the world's staple food is produced from rain-fed fields, along with meat production from grazing, and the production of wood from forestry (FAO, 1999). In sub-Saharan Africa, almost the entire food production, along with major industrial products such as cotton, tobacco, and wood, is produced from green water in rain-fed fields (Savenije, 1999). Water policy and planning focus almost exclusively on blue water, perhaps because green water can be managed only indirectly (FAO, 1999). This marginalization of rain-fed agriculture affects the accounting of water availability, since green water is not accounted for as water available for human use. Rain-fed agriculture in temperate zones tends to be highly mechanized and economically efficient, if energy demanding. In the tropics, small-scale farming is by far more prevalent. Small farms can also be run very efficiently, and in the wet tropics in countries such as Indonesia, Bangladesh, and Taiwan, there is often enough water to allow a second or third harvest through the use of irrigation (Savenjie, 1999). In the semi-arid tropics, especially, however, most agriculture is subsistence farming with very low efficiency. Africa, for example, gets at least 90 percent of its food from such farms. Food production could be increased in the area where it will be consumed by improving the efficiency of the use of green water for agriculture in these areas. In the semi-arid tropics, supplementary irrigation from groundwater can be used to optimize the use of green water for crop production. Relatively small quantities of blue water can safeguard crop production and, by doing so, produce high yields per cubic meter of blue water, much higher than can be attained from full-scale (dry season) irrigation (Savenije, 1999).

Groundwater Supplies

Approximately one-third of the world's population depends on groundwater supplies, some exclusively (WRI, 2000). Almost 20 percent of global water withdrawals are taken from groundwater supplies, mostly shallow aquifers (WRI, 2000). Runoff from precipitation recharges shallow aquifers, just as runoff feeds rivers, lakes, and streams. Deep aquifers tend to contain fossil water that is hydrologically disconnected from the surface hydrologic cycle (Revenga, Brunner, et al., 2000). There are problems associated with each of these categories of sources. Fossil groundwater is finite—it recharges only over extremely long periods, if at all, so that when it is withdrawn it is mined, and, once used, cannot easily be returned to the deep aquifer. Shallow aquifers, while rechargeable, are susceptible to pollution from agricultural and industrial chemicals, among other sources. They are intimately tied to freshwater ecosystems, so pollution or overdrafting (unsustainable withdrawals) of shallow aquifers also affects adjacent freshwater ecosystems, depriving them of a significant portion of their flow and polluting their waters.

Because many developing countries depend on groundwater for irrigation, and on irrigation for food supply, the uncontrolled depletion of groundwater aquifers is seen as a serious threat to food security in many areas, in particular the North China plains and western and peninsular India (IWMI, 2001). About two-thirds of unsustainable groundwater withdrawals occur in India, in areas that are responsible for over 25 percent of India's harvest (Shah, 2000). The consequences include saline ingress into coastal aquifers, arsenic contamination, fluoride contamination, as well as declining well yields and increased pumping costs (Shah, 2000). Other areas that are also in danger of depleting their groundwater resources are urban areas across Asia, parts of Mexico, the High Plains (or Oglalla) aquifer in the Midwestern United States, Yemen (where groundwater abstraction exceeds recharge for the country as a whole by 70 percent (Briscoe, 1999), Saudi Arabia (where 90 percent of water supplies are achieved through groundwater mining) (PAI, 1993), and other Middle and Near Eastern countries. It is difficult to know the true extent of groundwater depletion, since reliable data on resource extent and use are lacking to a greater degree even than that for surface water.

The Role of Freshwater Ecosystems

Natural freshwater ecosystems play an important role in determining water quality and quantity and the sustainability of water resources. They provide a range of critical, life-supporting functions, including the cleaning and recycling of the water itself. A freshwater ecosystem is a group of interacting organisms dependent on one another and their water environments for nutrients (e.g., nitrogen and phosphorus) and shelter. It includes, for example, the plants and animals living in a pond, the pond water, and all the substances dissolved or suspended in that water, together with the rocks, mud, and decaying matter at the bottom of the pond. Freshwater ecosystems can be running waters (such as rivers and streams), lakes, floodplains and wetlands, or freshwater or brackish coastal habitats. Freshwater ecosystems provide water supply for human consumption and agricultural cultivation. They provide a host of critical goods and services for humans and other living beings including water, food fish, energy, biodiversity, waste removal, and recreation. These ecosystems, in turn, are important components of the hydrological cycle.

In theory, natural ecosystems are self-sustaining. However, major changes to an ecosystem, such as climate change, land-use changes, or the removal of a species, may threaten the system's sustainability and result in its eventual destruction (Scientific American, 1999). Destruction of or damage to a freshwater ecosystem removes it from or impairs it in its role in the hydrological cycle and diminishes the amount of freshwater the earth's environment can provide. Once a freshwater ecosystem is destroyed, it is difficult or impossible to restore it to its natural state.

Natural Influences on Water Availability

Geographic Location

Water is not distributed evenly across land masses, and much of it is far from population centers. A large percentage of global water resources is available in places such as the Amazon basin, Canada, and

Alaska (Cosgrove and Rijsberman, 2000). About three-quarters of annual rainfall comes down in areas where less than one-third of the world's population lives (Gleick, 1996). Because precipitation is also temporally uneven, many people are unable to make use of the majority of the hydrologically available freshwater supply. Rainfall and river runoffs occur in very large amounts in very short time periods and are available for human use only if stored in aquifers, reservoirs, or tanks (Cosgrove and Rijsberman, 2000).

Figure 2.3 shows the large spatial variation in blue water availability, as well as the influence of population on overall availability. The two extreme cases are Asia and Australia/Oceania. Asia has the highest total water availability of the continents, but the least per-capita availability, while the opposite holds true for Australia/Oceania due to lower total population. Figure 2.4 shows the per-capita renewable water supply as calculated by river basin. The areas that have less aggregate per-capita water supply are more apparent, but additional spatial information gives a slightly clearer picture. For example, looking again at Australia, it is clear from Figure 2.4 that some areas, the most populated areas, as it turns out, have large per-capita supplies of water, while there are other parts of the continent, where very few people live, that have very small per-capita supplies of water. In Asia, it is apparent from this map that, although the continent as a whole may have relatively low per-capita water availability, some areas are quite well endowed with water. Again, the level of aggregation of water supply data makes a great deal of difference in the interpretation of the data.

Figure 2.3
Continental Total and Per-Capita Blue Water Availability

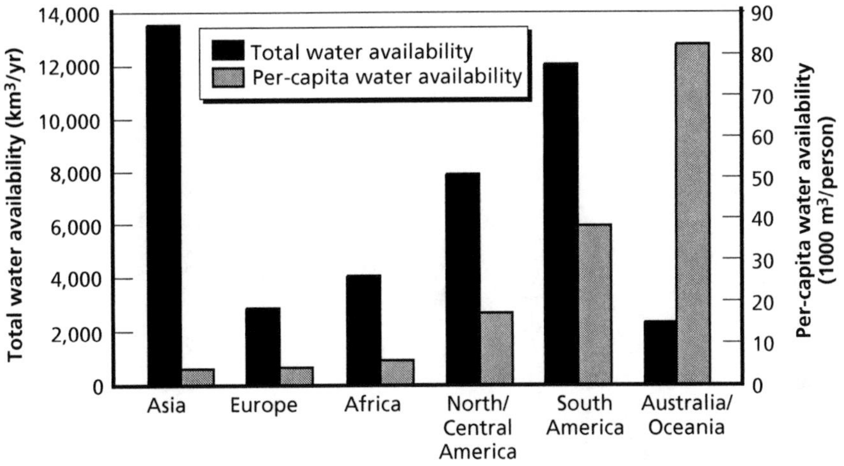

SOURCE: Shiklomanov (1998), in Gleick (2000a).
RAND *MG358-2.3*

Climate Change

Most researchers agree that climate change will result in higher variability of precipitation in many areas. Higher global and regional temperatures will mean increased evapotranspiration, changes in snowfall and rainfall patterns, and changes in the intensity, severity, and timing of major storms, among others (IUCN/WWF, 1998). This will result in changes in soil moisture, water runoff, and regional hydrology (El-Ashry, 1995). Arid and semi-arid regions will be most affected by these alterations (Frederick and Gleick, 1999). Because of these changes, the supply of and demand for water will change, as will its quality.

Changes will differ at various latitudes and will, like all weather and water resources, be locally specific. Some watersheds will experience increased runoff, more extreme storm activity, decreased soil moisture, increased groundwater recharge, or increased incidence of drought, or the opposite of these (IUCN/WWF, 1998). In some

Figure 2.4
Annual Renewable Water Supply Per Person by River Basin, 1995

locations, the changes will be beneficial in terms of water management; in others they will be detrimental. Ecosystems are similarly variably affected. The severity of the effects will depend on the type of change, the ecosystem itself, and the nature of any human interventions. Some of these effects may include changes in the mix of plant species in a region, lake and stream temperatures, thermocline depth and productivity, lake levels, mixing regimes, water residence times, water clarity, reduced extent of wetlands, extinction of endemic fish species, exotic species invasions, and altered food web structure (IUCN/WWF, 1998).

Deep aquifers will be less affected and may become even more important as a source of fresh water (Rivera, Allen, and Maathuis, 2003). Shallow groundwater may be affected by climate change—not only the shallow aquifers that may be depleted or the cause of waterlogging by changing recharge patterns, but also coastal aquifers and wells experiencing saltwater contamination through rising sea level.

Another potential repercussion of climate change is the warming of the earth's lakes. Lake temperatures may rise as much as seven degrees Celsius under carbon-dioxide doubling, decreasing the lakes' ability to naturally cleanse themselves of pollution (Poff, Brinson, and Day, 2002, pp. 13–15). Additional concern comes from the fact that these changes will evolve in a relatively short time frame. This will leave little room for adaptation or coping by the ecosystems, increasing the likelihood that they will be negatively impacted.

Climate change will have indirect effects on water resources as well, such as hydroelectric generation, human health, navigation and shipping, agriculture, and water quality (IUCN/WWF, 1998). For example, changed temperatures, flows, runoff rates and timing, and the ability of freshwater ecosystems to assimilate wastes may all affect water quality (IUCN/WWF, 1998). Many of these effects may be mitigated or worsened by human actions, so it is important to plan and take into account such changes to the extent that they can be recognized and predicted. Climate change may exacerbate changes in patterns of demand, population distribution, and poverty and food insecurity in marginalized communities. For example, arid and semiarid areas that receive less rain or rain in fewer, more concentrated time periods may experience flooding, increased crop failures, failure of wells, or other consequences that lead to out-migration or the need for water to be imported.

Calculating Water Availability

Hydrological water availability refers simply to the amount of water as liquid, gas, or solid, but not as molecularly associated with minerals and living resources in the earth's environment. However, the quantity of water practically available for humans is different from the hydrological measure of availability. The amount of water available to a population is defined by factors such as quality of water demanded, cost, environmental effects, political and legal agreements, individual wealth, and the technical ability to move water from place to place (Simmons, 1991; Gleick, 2000b). It also depends on the use to which the water will be put, because drinking water, for example, demands a higher quality of water than do industrial or recreational uses (Swanson, Doble, and Olsen, 1999).

Although we can report continental average and per-capita water availability with some confidence, as in Figure 2.3, the actual amount available to any person or group for a particular use is less certain. Though this would clearly be a more useful and meaningful measure to have, it is very difficult to calculate, especially in areas where water is not metered, or is not closely monitored. It is also dependent not only on physical availability, but also management and investment levels that result in the development of reservoirs, dams, and other technologies to capture runoff and groundwater (Bernstein, 2002).

As will be described in more detail later, water resources are best measured in units of watersheds and aquifers. When water availability is calculated for populations in individual river basins (not including fossil water sources), 41 percent of the world's population (approximately 2.3 billion people) was found to live in river basins where per-capita water availability was less than 1,700 m^3 per year. This is the level used as a cutoff by many water experts to indicate water-stressed populations (WRI, 2000). About 1.6 billion people, or 19 percent of the world's population, are found in this analysis to live in conditions of water scarcity, defined as water availability of less than 1,000 m^3 per person per year (WRI, 2000). A different conclusion can be drawn when national boundaries are used as units. Earlier estimates using national water balances found 11 percent of the global population to be water stressed. As touched on earlier, the definitions of stress or scarcity used here are problematic under any calculations of water availability. The extreme differences in water availability that are created by the two different methods of calculation point out the dangers of putting too much credence in any such calculations. As discussed earlier, even the more sophisticated estimates based on watersheds are deceiving. They hide variability of access, temporal and spatial availability, and legal and political restrictions, and are therefore far from the ideal of individual water availability measures. Most importantly, they may over- or underemphasize existing or potential signs of stress, and hence poorly constrain the choices faced by policymakers, water managers, and others.

CHAPTER THREE

Patterns of Demand for Fresh Water

Water use is commonly analyzed according to three categories or sectors: agricultural, industrial, and domestic or municipal.[1] Water utilization in each sector is influenced by many factors, including population variables such as growth, distribution, and composition. This chapter explains how changes in each sector affect the demand for freshwater by a given population.

Demand varies geographically. There are many regional differences in the demand for water. Agricultural demand ranges from zero percent of total freshwater withdrawals in the Maldives to 99 percent in Afghanistan, Nepal, Guyana, and Madagascar. Industrial use ranges from a low of zero percent in several countries such as Burundi, Belize, and Somalia, to a high of 85 percent in Belgium. Finally, the domestic sector withdrawal of water ranges from one percent in Turkmenistan, Afghanistan, Nepal, Guyana, and Madagascar, to 87 percent in Malta. These broad differences emphasize the importance of looking at water withdrawals on as small a scale as possible, since higher levels of aggregation can give a false sense of what is actually taking place.

Demand varies over time. Demand by category varies by country and changes over time, as is demonstrated by the history of water withdrawals in the United States in Figure 3.1. By 1995, agricultural demand for water accounted for 42 percent of total U.S. demand, with industry consuming 46 percent and the domestic sector 12 per-

[1] Sometimes a fourth sector, thermoelectric power, is also assessed.

cent (Gleick, 2000b). This came within the context of an overall decline in water use, in the U.S. example. Worldwide, agriculture is the most demanding sector, accounting for about 70 percent of all annual water withdrawals, as shown in Figure 3.2. Industry draws about 20 percent and domestic use about 10 percent (FAO, 2003a).

Demand varies in relative quantity. Despite large variability within regions, regional figures can give clues as to the cause of differences in water use between locales, as seen in Figure 3.3. First, there are significant regional differences in the relative quantity of sectoral water withdrawals. In all regions except Africa and South America, the domestic sector withdraws the least water, proportionally. In Africa and South America, economies are generally agriculturally based and in many places, standards of living are low. In contrast, Europe, a heavily industrialized region with a high standard of living, uses

Figure 3.1
U.S. Water Use 1900–1995 by End-Use Sector (Billions of Gallons Per Day)

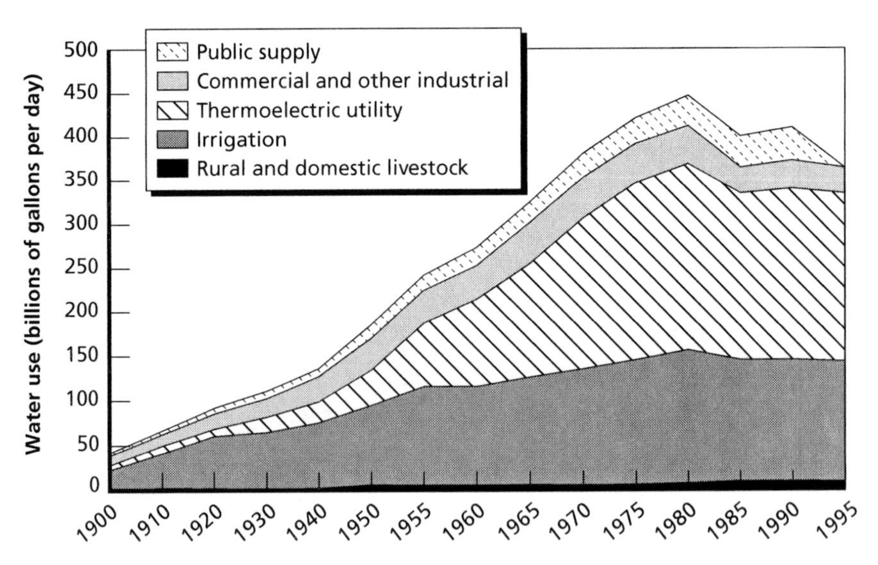

SOURCE: CEQ (1997).
RAND *MG358-3.1*

nearly half of its water withdrawals for industry, while agriculture accounts for only about a third. In two other economically developed regions, North America and Australia/Oceania, industry withdraws a large share of water, but still less than agriculture. These regions are, while industrialized, also very agriculturally active. Asia is remarkable for the fact that fully 90 percent of its water withdrawals went to agriculture in 1995. This reflects its heavy reliance on irrigation in agriculture and its still-developing industrial sector.

Demand varies in its seasonal timing. The timing of water demand differs among sectors as well. Agricultural water use tends to be seasonal, though irrigation can create year-round farming in some areas. Domestic and industrial demands are year round, which means that in climates with seasonal rainfall or runoff there must be a form of storage in order to have dry-season supply (Meinzen-Dick and Appasamy, 2002). Groundwater or surface reservoirs are the most common form of storage.

Figure 3.2
Worldwide Sectoral Water Use

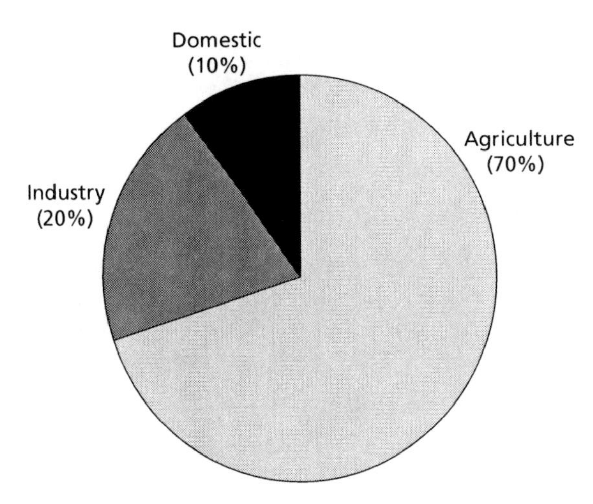

SOURCE: FAO (2003a).
RAND *MG358-3.2*

Figure 3.3
Estimated Sectoral Water Withdrawals, by Region, 1995 (Percent of Annual Total)

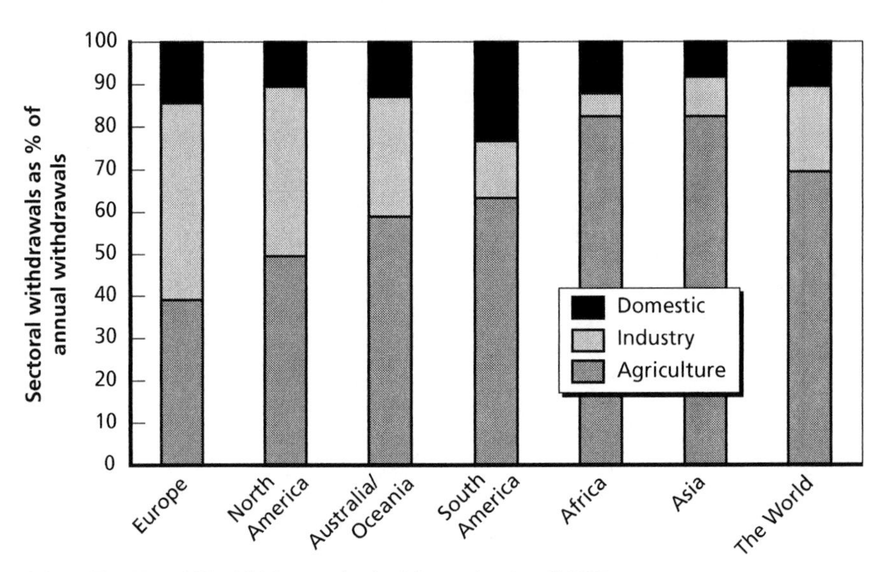

SOURCE: UN and World Meteorological Organization (1997).
RAND *MG358-3.3*

Agricultural Sector

Agriculture is generally the largest user of freshwater, though it ranks behind industry in Europe. Most, though far from all, water that is withdrawn for use in agriculture goes to irrigation. Losses from leakage and evaporation in most irrigation schemes are well over 50 percent, including those in the western United States (Falkenmark and Widstrand, 1992). Israel is often held up as an example of a nation with efficient irrigation, where the use of drip irrigation and other technologies and management practices keep their irrigation losses to as little as five percent (*Economist*, 1999). Because of the sparse and variable rainfall in semi-arid and arid areas, more irrigation is needed, so many countries in those regions have higher absolute and percentage withdrawals for agriculture than others in their economic class.

Globally, about 73 percent of water withdrawals in arid and semi-arid regions are for irrigation, versus around 60 percent for all areas (El-Ashry, 1995).

The green revolution, which increased food production in some developing areas from the 1960s through the 1980s, also encouraged the rapid development of irrigation systems in those countries, especially in Asia. Irrigated production provides 60 percent of the food in India and 70 percent of the grain in China (ECSP, 2002), and India uses 92 percent of all water for agricultural purposes, with just 5 percent for industry, and 3 percent for domestic use (Gleick, 2000b). Partly because of this trend, the area of land under irrigation is estimated to have increased more than six times in the last century (WRI, 1996). Half of the land currently under irrigation was added since 1950, but per-capita irrigated land is now decreasing (de Villiers, 2000). Rain-fed agriculture accounts for 60 percent of food production in developing countries on 80 percent of arable land (FAO, 2003b). Only 20 percent of the arable land in developing countries is irrigated, but it produces around 40 percent of all crops and close to 60 percent of cereal production. Much of irrigated farming is done by larger farmers on the best land, but the majority of farmers in developing countries rely on rain. Without irrigation, yields in the world's major breadbaskets would be nearly half of their current production (de Villiers, 2000).

Much of the future food production in developing countries is expected to come from irrigated land (FAO, 2003b). This is despite the fact that the massive investments in large public irrigation systems have performed below expectations on average, with crop yields and water-use efficiency below projections, and well below what is reasonably achievable (El-Ashry, 1995). Projections for 2030 for agricultural water withdrawals show a 14-percent increase (FAO, 2003b), although some have predicted increases in demand as great as 50 percent for irrigation water by 2025 (WRI, 2000). Much of the demand for irrigation is being met in nonsustainable ways (e.g., by pumping nonrenewable groundwater supplies) and therefore may be problematic in the future (El-Ashry, 1995). Because many developing countries depend on groundwater for irrigation, and on irrigation for food supply, the uncontrolled depletion of groundwater aquifers is seen as a serious threat to food security in many areas, in particular the North China plains and western and peninsular India (IWMI, 2001). Increases in production must be met by less water-intensive means. Irrigated cropland in dry areas with severe water limitations supported 40 percent of the world's population in 1985 (Vörösmarty et al., 2000). Issues of balancing water supply and demand are a concern, especially in arid and semi-arid regions, where 20 percent of recent world population growth has occurred (El-Ashry, 1995).

Industrial Sector

Industrial water use is usually the next largest category of user. Water is used in industry for cooling, cleaning, processing, generating steam power, and transportation. Some industries are particularly water intensive, such as food processing, thermoelectricity production, and the paper and pulp industry. However, in any industry there are efficient and inefficient examples of water use, depending on management practices and technology used. In the United States, industry used 40 percent less water in 1995 than in 1970, both by increasing efficiency and because of changes in the mix of industries there (Gleick, 2000a). Similar improvements in water-use efficiency occurred in many other industrialized countries as well in the 1970s and 1980s. Britain and Denmark, for example, have reduced their water usage overall by more than 20 percent since 1980 (*Economist*, 1999). There is a significant opportunity to reuse water within factories, which could add significantly to the efficiency of water use in that sector. Because industries tend to cluster in urban areas, industrial water withdrawals are a significant component of urban water demand. There are a growing number of industries in rural areas, however, adding to rural demand.

Domestic and Municipal Sector

Water use in the domestic and municipal sector is generally the smallest share of water use, except in countries with little agriculture or industry such as Lithuania, Equatorial Guinea, the Maldives, and Malta. Water is used in the domestic and municipal sector for personal uses such as drinking, in food preparation, for sanitation purposes, for watering landscapes, for maintaining swimming pools, and other such applications. The absolute amount of water withdrawn for personal and municipal uses varies widely, from a minimum of about one or two cubic meters per person per year in low-income countries such as Cambodia, Haiti, the Gambia, Somalia, or Mali, to about 245 m^3 per person per year in New Zealand, 240 m^3 per person per year in Armenia, and over 200 m^3 per person per year in the United Arab Emirates and the United States. In the case of some low-income countries, water use in the domestic and municipal sector is expected to increase in the future, since there are supply problems that result in high levels of unmet demand. In countries that are using much higher absolute and per-capita levels of water, there are many chances for increasing efficiencies and reducing total use. A number of cities in both low- and high-income countries such as Jerusalem, Israel; Mexico City, Mexico; Los Angeles, California; Beijing, China; Singapore; Boston, Massachusetts; Waterloo, Canada; Bogor, Indonesia; and Melbourne, Australia, have reduced water use from 10 to 30 percent through municipal conservation programs (Postel, 1997). Domestic and municipal sector water use is centered in urban areas, and can dominate urban water demand, depending on the amount of industry located in a particular urban area.

The domestic sector, although less important as a source of demand than other sectors in most places, warrants special attention because of its implications for health and mortality. Poor water quality is a major factor in infant mortality, premature mortality, and lost productivity in many countries, especially in the developing world. The gastrointestinal disorders associated with poor water quality are epidemic in some areas. Safe drinking water is a matter of great public health, and political concern. This subject will be discussed at greater length later in this monograph.

The sectoral demands for water are an important determinant of how demographic factors will affect future demand. Although the domestic sector is often most connected to demographic factors in discussions on water resources, in general other sectors have more impact on demand for water. It is important to investigate how these sectors and demographic factors interact, as well as to ascertain which sectors are most important to the locale or region under investigation.

CHAPTER FOUR

Demographic Influences on Water Resources

Demographic factors affect consumption of water and the quality and health of natural ecosystems both directly and indirectly. As discussed earlier, population growth is usually hypothesized as one of the most important factors threatening the sustainability of water systems. Urbanization, migration, and number of households are demographic factors that are less commonly associated with demographic effects on water resources, although they are also important. Other factors that affect water use tend to be indirectly linked to population—income levels, a rise in living standards, modifications to landscapes and land use, contamination of water supplies, and inefficiency of water use caused by a failure to manage demand (Winpenny, undated). The following sections will discuss the direct and indirect population factors that affect human utilization of water resources. They also review the main demographic changes that have occurred globally.

Population Size

Population size is fundamentally linked to water use. Although the relationship is nonlinear, in general the more people there are, the more resources, such as freshwater, humans will ultimately consume. The relationship is not simply linear because different populations and individuals will use resources in different ways and in different amounts due to mediating factors and resource limitations. Nonetheless, population size and changes in it are very important factors in the magnitude of water use and of the human impact on the freshwater ecosystems that supply the water. Population growth is perceived by some to be the most important demographic trend affecting water resources (Hinrichsen, Robey, and Upadhyay, 1997).

Population distribution and trends can affect water quality. Humans compromise the quality of water through some types of changes in landscapes. Land-use activities in a watershed can affect the quality of surface water as contaminants are carried by runoff and deposited in streams and rivers. Land-use activities also affect groundwater, especially through infiltration of pollutants left over from agricultural, industrial, commercial, and other human activity. Soil itself becomes a water pollutant as it is eroded from farmlands. Water withdrawals can similarly be detrimental to water quality by reducing the flow of water through the system, thereby concentrating both naturally occurring and human-induced contaminants.

The 20th century saw world population grow from 1.65 billion to six billion people, with almost 80 percent of that increase taking place after 1950 (Figure 4.1). This rapid growth was sparked by mortality reductions, partly due to improvements in water quality. This was particularly true in developing countries, where average life expectancy at birth increased by more than 20 years over the last half of the century, from 41 years to 64 years (UN, 2001). World population has increased by nearly two and a half times since 1950, with the last billion people added in just 12 years (1987–1999), the fastest growth in history.

However, world fertility decline since the 1970s has meant a significant decrease in the global rate of population growth (see Figures 4.2 and 4.3). Growth rates have dropped to a low of about 1.2 percent per year for 2000 to 2005, adding about 77 million people annually (UN, 2001). The average world total fertility rate has dropped from 4.9 children per woman in 1965 to 1970 to 2.7 in 2000 to 2005, a decline of 45 percent. The total fertility rate has declined markedly in both developed and developing countries over this period, from 6.0 to 2.9 children per woman in developing regions and from 2.4 to 1.5 children per woman in more developed regions.

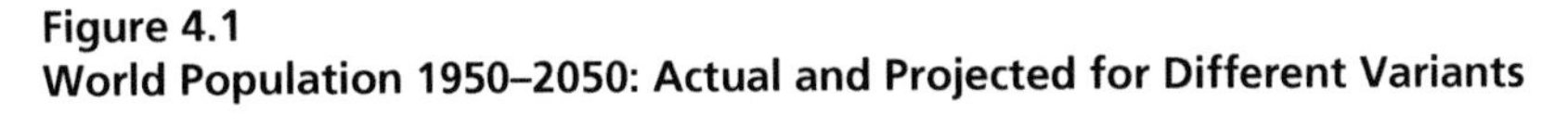

Figure 4.1
World Population 1950–2050: Actual and Projected for Different Variants

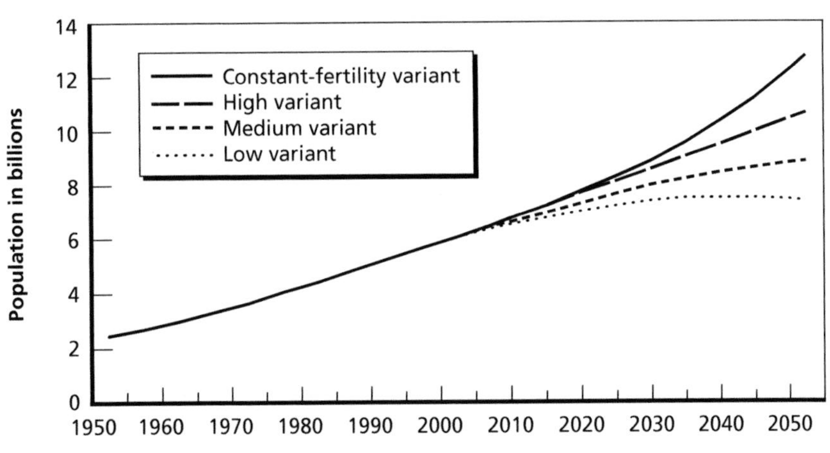

SOURCE: UN (2002b).
RAND *MG358-4.1*

Despite these fertility rate declines, global population is still increasing, especially in the developing regions, as projected in Figure 4.4. While slowing, this population momentum will continue for some time due to the large number of women of reproductive age, a result of fertility rates in the past. Because the population is sensitive to sustained changes in fertility levels, the population projections for 2050 range from less than eight billion to almost 11 billion (Figure 4.1). This sensitivity was highlighted when the United Nations revised downward its 1996 population projection in 2000 due to unexpected decreases in fertility levels.

Figure 4.2
Population Growth Rates, Actual and Projected, 1950–2055

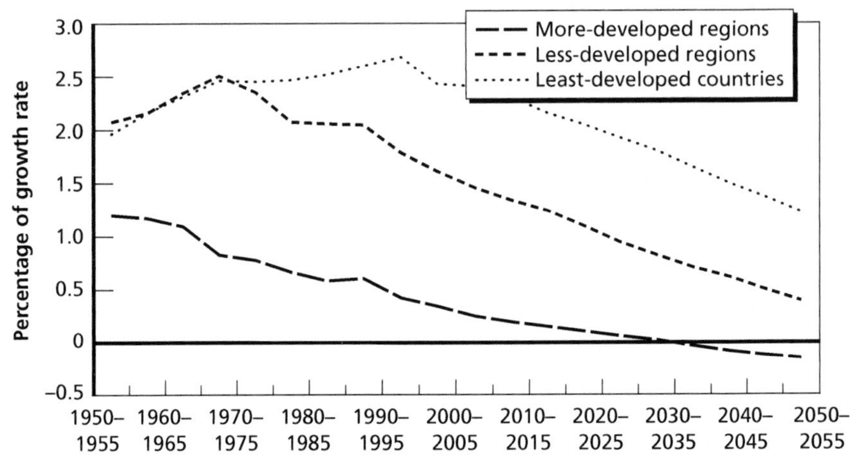

SOURCE: UN (2002b, 2003).
RAND *MG358-4.2*

Figure 4.3
Total Fertility Rates: Actual and Projected, 1950–2055 (Children Per Woman)

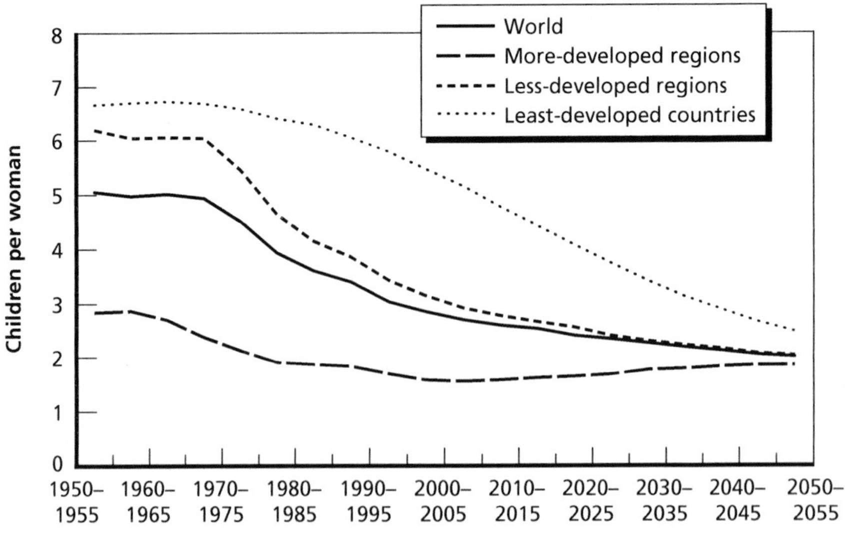

SOURCE: UN (2002b, 2003).
RAND *MG358-4.3*

Figure 4.4
World Population 1950–2050, Medium Variant, by Economic Status of Countries

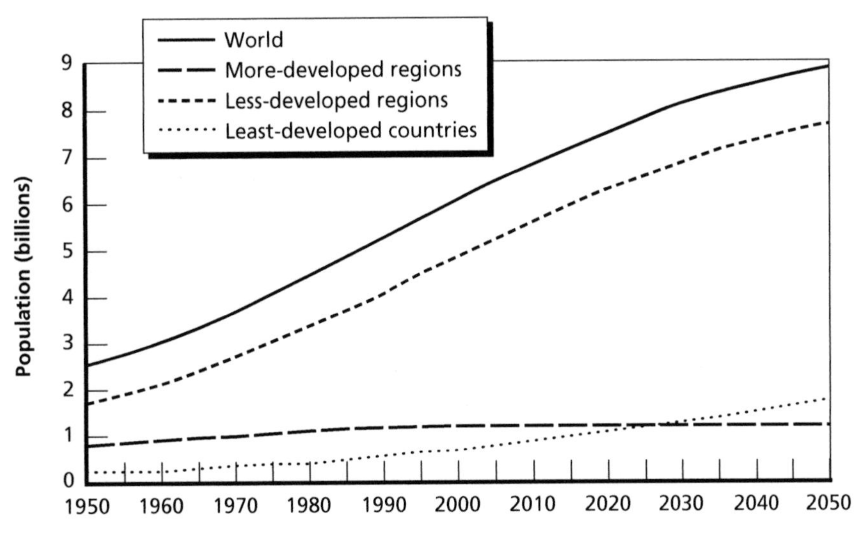

SOURCE: UN (2002b, 2003).
RAND *MG358-4.4*

Number of Households

The number of households is increasing worldwide, as the average household size is decreasing. It seems likely that the decrease in household size will affect future domestic water demand, due both to household-related economies of scale and to a decrease in the effectiveness of investments in technical water-saving measures (Martin, 1999). One study looking at countries of the Middle East found that projections of domestic water demand based on a per-household approach resulted in up to 75-percent higher estimates than projections based on a per-person approach (Martin, 1999). In Australia, a study found that the household water budget (milliliters per capita) decreases with increasing household size, as on average, households' water use in Sydney ranged from about 1.1 ml per person for a one-person household to 0.5 ml per person for a seven-person household

(Lenzen, 2002). This implies that decreasing household size may have a significant impact on future demand for water in the domestic sector, and even that this impact, in some places, could be more significant than simple population growth.

Lawn watering, house cleaning, car washing, laundry, and dishwashing are all examples of water uses that household-related economies of scale may affect (a six-person household will use less than twice the water for these tasks as a three-person household). Water-saving technology investments that are reduced in cost effectiveness by decreases in household size include water-saving toilets and showerheads, each of which will cut water consumption twice as much in a six-person household as in a three-person household.

Additionally, a growth in the number of households will have indirect effects on water supply and quality. Compared to a household of four persons or more, which exploit economies of scale, one- to three-person households consume more water and produce more waste per capita. An increase in numbers of households fueled by a transition to smaller household size requires more housing units, increasing the area covered by housing and the materials needed for construction. This generally is associated with an increase in urban sprawl, which also affects water resources and water quality negatively. Sprawl development paves over land that would normally allow water to seep into aquifers and replenish lakes and rivers, but instead allows it to run unfiltered and polluted directly into those same bodies of water (Shogren, 2002).

This increase is due to a combination of simple population growth and behavioral changes, which are creating additional households within the same population. Some studies have found evidence that an increase in households containing fewer people does more environmental damage than simple population growth (Liu et al., 2003; MacKellar et al., 1995). In one, the researchers looked at 141 countries, including 76 countries containing biodiversity hotspots, such as Australia, India, Kenya, Brazil, China, Italy, and the United States. The growth in the number of new housing units worldwide increased at a rate more rapid than the population growth—particularly in these countries. Between 1985 and 2000, the number of households grew by 3.1 percent a year, while the population grew only 1.8 percent in these countries (Liu et al, 2003). Another study found that between 1970 and 1990, the growth in the number of households in developed regions had more than double the impact on growth in CO^2 emissions than did the growth in population numbers (MacKellar et al., 1995).

Even in some European countries with negative total population growth, the number of households is increasing. Between 1980 and 1995, the urban population in Western Europe increased by nine percent (UN, 2001) but the number of households in the area increased by 19 percent (EEA, 2001). In North America, household growth rates of 1.6 percent in 1995 exceeded total and urban population growth rates of 1.1 percent and 1.0 percent respectively (UN Human Settlements Programme, 1999). Over the same period, the household size in hotspot countries fell from 4.7 to 4.0, primarily because of lower fertility rates, higher divorce rates, higher per-capita income, aging populations (resulting in more one or two person households), and a decline in multigenerational family units.

Urbanization and Migration

Urbanization and migration are other important population trends that affect water resources. Urbanization, both by degree and rate of growth, affects the level of water use within a country. It can affect the levels of per-capita use, overtax water resources by concentrating demand in a small area, and overwhelm existing infrastructure. The redistribution of population by migration can shift pressures on water resources, primarily as a major contributor to urbanization.

The world is rapidly becoming more urban, as the total population living in urban areas is expected to equal that living in rural areas by about 2007 (Figure 4.5). This compares with a 30-percent world urbanization level in 1950. By 2030, every major region in the world will be majority urban, with expectations that 84 percent of people in the developed world and 56 percent in the less-developed world will

Figure 4.5
World Urbanization Trends: Actual and Projected

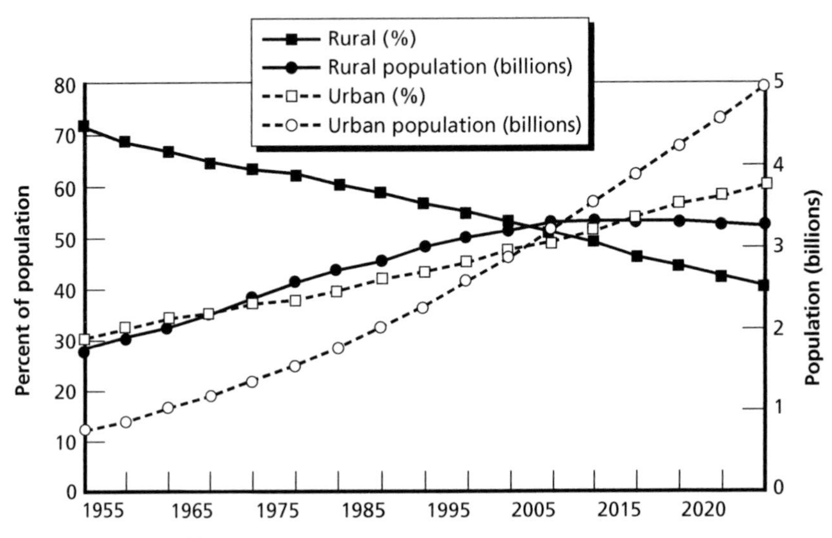

SOURCE: UN (2002b).
RAND *MG358-4.5*

live in urban areas (UN, 2001). In fact, virtually all the population growth expected at the world level between 2000 and 2030 will be concentrated in urban areas. The world's urban population reached 2.9 billion in 2000 and is expected to rise to 5 billion by 2030. Population growth will be particularly rapid in the urban areas of less-developed regions, averaging 2.4 percent per year between 2000 and 2030. This is caused both by natural increase (more births than deaths) and by high levels of rural-urban migration and the transformation of rural settlements into cities.

There are marked differences in the level and pace of urbanization among areas in the less-developed regions of the world. For example, Latin America and the Caribbean are highly urbanized, with 75 percent of the population living in urban settlements in 2000, a proportion higher than that of Europe. By 2030, 84 percent of the population of Latin America and the Caribbean are expected to be urban, a level similar to that of Northern America, the most highly urbanized area of the world by 2030. Africa and Asia are considerably less urbanized, with 37 percent of their respective populations living in urban areas in 2000. They are expected to experience rapid rates of urbanization so that by 2030, about 53 percent of their inhabitants will live in urban areas (UN, 2002b). The rural populations of the world are expected to begin to decline in absolute numbers by 2020.

Migration is an important factor in the high rates of urban growth in many countries, especially in the developing world (Gardner and Blackburn, 1996). The United Nations estimated that over 50 percent of urban growth in the 1980s was due to net rural-to-urban migration and the reclassification of rural areas as urban, largely because 72 percent of urban growth in China during the 1980s was attributable to migration and reclassification (UN, 1996). In Africa, migration from rural areas accounted for about 25 percent of urban growth in the 1980s and 1990s; in Asia, excluding China, over 45 percent of urban growth in the 1980s has been due to migration and reclassification of territory from rural to urban; and in Latin America the proportion is about 35 percent (UN, 1996). Attention has mostly been paid to rural-urban migration, despite the fact that rural-to-rural and urban-to-urban migrations are more common in many countries (National Council for Science and the Environment, 1994).

International migration is also a factor in urbanization. Including refugees and undocumented migrants, about two percent of the global population is currently living outside the countries of their birth. The more developed regions are expected to remain net receivers of international migrants, with an average gain of about two million per year over the next 50 years, about two thirds of whom will originate in developing countries (UN, 2001).

Urbanization affects the level of water use within a country. This is particularly true for the domestic and municipal sector, where urbanization—and the infrastructure that often accompanies it—can make a significant difference in per-capita use. It is important to keep in mind that domestic water use in many countries is a relatively small part of the freshwater demand burden. However, the densities of urban areas can mean that local demand can be extremely high, outstripping the resources available locally, creating environmental havoc and localized water shortages, or impact more widely when resources are transferred from elsewhere. About 60 percent of all freshwater withdrawals are used directly or indirectly in urban areas—directly in industry and domestic drinking water and sanitation, and indirectly through the consumption of irrigated crops (Sheehan, 1999). The food consumed by cities may be obtained from either nearby or distant farms, meaning that the impact of the city on water resources may stretch far beyond its borders.

Urbanization of populations can affect water quality when formerly vegetation-covered land is changed to pavement and buildings, increasing the volume of runoff and its pollution levels, degrading or eliminating the ability of the land to absorb rainwater, and possibly putting human wastes into water systems. If urbanization and population growth requires increased agricultural land, water quality may be degraded from any changes in land use (i.e., conversion from forest to agricultural use) and from the wastes of the agriculture practiced on the land. Cherapunji, in northeast India, has seen extensive land clearing, specifically deforestation, contribute to a state of water scarcity in one of the wettest places on earth (FAO, 1999; Spaeth, 1996).

A recent study has concluded that conserving forests is the best and most cost-efficient way to secure water supplies around cities (Dudley and Stolton, 2003). Forests naturally reduce erosion, sediment, and pollutants, and often store water. In one case, New York City decided to meet its increasing water needs by reducing its demand for water, while at the same time protecting the abundant, clean, and inexpensive, but far-flung, drinking water supply system (Platt, 2001). The plan, which included increased water taxes and payments to farmers to reduce fertilizer use and grazing, resulted in a strengthened agricultural presence in the watershed region, improved management practices by the farmers, and improved water quality (UN, 1998). It also cost about $1 billion, which is $6 billion less than that required to build a new water treatment plant, representing a substantial savings.

The problems of meeting the demands of urban sanitation and domestic freshwater are made more difficult by the demographic realities of urban growth. This is especially true in places where population growth is occurring fastest, in small cities and informal settlements surrounding large cities, as is happening in much of the developing world (Bernstein, 2002). In developing countries, this rapid urban growth often puts tremendous pressure on antiquated, inadequate water supply systems. Sanitation and piped water supply are often overwhelmed or nonexistent (IUCN, 2000). In Africa, 27 percent of the urban population live in informal settlements, areas of minimal infrastructural improvements, while in Asia, 18 percent, and in Latin America and the Caribbean, nine percent of the population live in such settlements (WHO/UNICEF Joint Monitoring Programme for Water Supply and Sanitation, 2000). These problems are compounded by inefficient delivery mechanisms, lack of financial resources, and insufficient management and technical skills (IUCN, 2000). Many agencies in lower-income countries are not equipped to manage the urban water supply, while some countries have ineffective water allocation systems that allow cities to run short of water at the same time that water resources are being used for subsidized agriculture. Because population in cities is growing faster than service provision, it is highly likely that percentages of populations served with clean water supplies and sanitation services in cities will decline in the future, leaving larger numbers of people with lower quality of life, and impacting local environments with untreated waste. This problem will be discussed in more detail in the following chapter.

Water use is increased when it is easy to obtain. The increasing impact of urban water use on water resources is not due only to absolute increases in urban populations. The ease of obtaining water is crucial to the amount used by an individual and, to a lesser extent, an industry or agriculturalist. Both piped water systems and water-based sewage systems are characteristics of urban systems that greatly increase water use. For example, in 1900 the average American household used as little as 10 cubic meters of water per year compared with more than 200 cubic meters today (Buckley, 1994). In 1900, most Americans obtained their freshwater from wells or public standpipes. Today, virtually every American household has running water available, and the water costs its users very little. This connection between easy access to water and increased water withdrawals is repeated in many countries with limited water infrastructure today. The low levels of household water use in low-income countries as compared to industrialized countries is partly a reflection of the difficulty many households in developing countries have in obtaining water.

Waterborne sanitation systems are a major factor in the higher rates of freshwater withdrawals in urban areas of developing countries and throughout industrialized countries. In modern India, for example, average use of water in urban areas ranges from 40 to 125 liters per person per day depending on availability of piped water supply and underground sewage (ECSP, 2002). This threefold difference, multiplied by large populations in urban areas, can make a large difference to a country or region's overall water withdrawals. At present, access to waterborne sanitation systems is limited mostly to people in the urban areas of most countries. In these systems, water is used to flush waste and other used water into a sewage system, which then must be collected and treated. Current wastewater systems use an average of 40 liters of pure water per day per person just for flushing, a water-intensive, energy-intensive, and expensive method for dealing with wastes (Gumbo, 2001; Austin and van Vuuren, 2002).

Groundwater pollution, nutrient pollution, and agricultural intensity are correlated with population density after a certain threshold is reached (Squillace, Scott, et al., 2002). The higher the population density, the greater the detection frequency of volatile organic compounds (VOCs) in aquifers, to the extent that untreated groundwater in urban areas is four times more likely to exceed a drinking-water criterion that untreated water in rural areas (Squillace, Moran, et al., 1999).

Fertility rates may be linked to resource access. Several studies have found that water (or fuel wood) scarcity leads to increased fertility in rural areas of the developing world. In rural, agricultural regions, wood and water are often collected as a common resource from the surrounding areas. It is generally the task of women and children to gather these resources, and as these common-property resources become scarce, each additional child provides a marginal benefit to the family through his or her labor (Dasgupta, 2000; Sutherland, Carr, and Curtis, 2003). Evidence for this impact has been seen in cross-sectional studies in Pakistan, Nepal, and South Africa (Biddlecom, Axinn, and Barber, 2005; Aggarwal, Netanyahu, and Romano, 2001; Filmer and Pritchett, 2002). In arid regions of the world, the same phenomenon is encountered, since resources, especially water, are by definition relatively scarce. In many agriculturally based rural or arid areas, therefore, water scarcity aggravates demographic variables, counterintuitively increasing population size (assuming child survival) and therefore increasing the potential resource use.

Economic Development

The level of economic development experienced by a country or region is a set of mitigating factors strongly tied to demographic factors. As sectoral analyses of freshwater withdrawals imply, the level of freshwater use is tied to a country's level of economic development. The correlation has been held as important enough in the past so that water use is often used as one of many measures of economic development (Rock, 2000). Overall, developing countries use less water per capita than do industrialized countries, but there is much variation within that correlation. Because freshwater is used to run industry, grow food, generate electricity, and carry pollution and wastes, it seems logical that increased industrialization or development leads to increased water use per capita. Both the individual water-use levels and the water used by society to support individuals and their ways of life are higher in industrialized countries. For example, as living standards rise, eating habits often also change. As incomes rise, the consumption of livestock products, such as meat and dairy products, as well as processed foods increases (Regmi and Dyck, 2001). Additionally, urbanization and rising incomes are often accompanied by changes from root crops and coarse grains to rice and wheat, as women, in particular, are attracted to foods with shorter preparation times. These changes imply higher per-capita demand for water, since the foods that are preferred by higher-income or urbanized residents tend to be significantly more water intensive in nature.

Per-capita freshwater use is higher in high-income countries. Typically, daily domestic water use by people in developed countries such as the United Kingdom and the United States averages 334 to 578 liters per person, while the average in Africa is 47 liters per person, and in Asia, 95 liters per person per day is the average (UNFPA, 2002). North American cities use, on average, 400 liters per capita per day while Western European cities use an average of 200 liters per capita per day (Cosgrove and Rijsberman, 2000). But since domestic and municipal water withdrawals are generally less voluminous in overall water withdrawals than agriculture and industry, even profligate use in the domestic sector tends to be overshadowed.

Overall per-capita freshwater withdrawal rates are also higher in industrialized countries. Based on average per-capita withdrawal rates calculated by World Bank income categories, low-income countries use about 320 cubic meters of water per person, lower middle-income countries about 370 cubic meters of water per person, upper middle-income countries 570 cubic meters of water per person, and high-income countries about 630 cubic meters of water per person, as is shown in Figure 4.6.[1] These data suggest that water use per capita goes up slightly as countries move from the low-income to lower middle-income group, and then experiences a significant increase as country income rises from lower to middle income. The withdrawal rates then again rise only slightly between countries in the upper–middle- and upper-income categories. This suggests that there may be structural economic changes or behavioral changes that occur as countries move from lower- to middle-income status that increase amount of water used in the country.

---

[1] There is a large amount of variance in these averages. A group of Central Asian countries (Azerbaijan, Kazakhstan, Kyrgyzstan, Tajikistan, Turkmenistan, and Uzbekistan) whose water usage is much higher than others in their income category have been removed from the data set. Their unique history as part of the former Soviet Union may influence their economies in ways that will change over time to become more in line with others in their income category. When this set of countries is included, average water withdrawal is 447 m^3 per capita in low-income countries and 821 m^3 per capita in lower–middle-income countries.

Figure 4.6
Per-Capita Water Withdrawals by Country Income

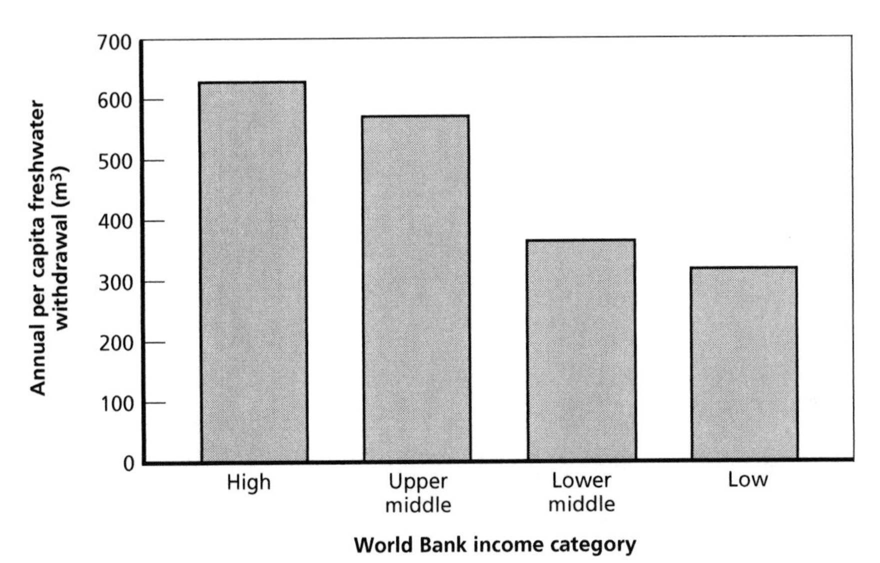

SOURCE: WRI (2000); World Bank (2002).
RAND *MG358-4.6*

Per-capita withdrawals also show variation by region. Withdrawals per person in North America and Europe are greater than in regions with less industrialized economies, though in Figure 4.7 European withdrawals appear lower due to the inclusion of less-affluent Central Asian economies in the regional breakdown. Again, this regional breakdown hides a large amount of variability, both in economic level and in water abstraction rates.

Figure 4.7
Freshwater Withdrawals by Region

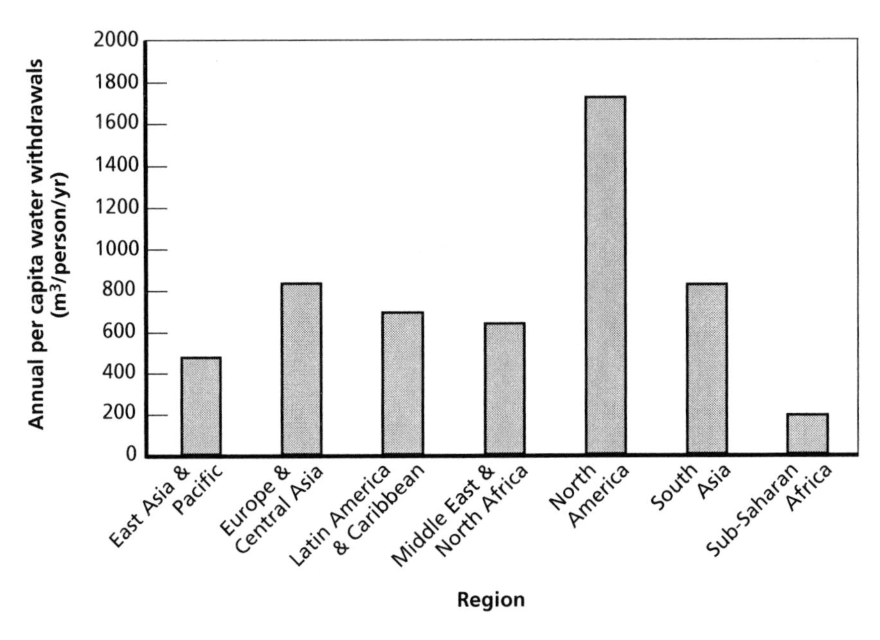

SOURCE: World Bank (2002).
RAND *MG358-4.7*

Some recent research has found that water-use intensity (water withdrawals per dollar of gross national product, or GNP) declines across countries as per-capita income increases (Rock, 2000). Water-use intensity can be seen as a proxy for the efficient use of water, in an economic sense. As a country develops, its water-use intensity may drop, probably because of structural changes in the economy as income is generated more from services and industry and less from agriculture. This is despite the fact that domestic and municipal use goes up with increases in income. There is often a period in the early industrial development of a country where the scale effects of economic expansion overcome the declining intensity of use effects so that overall and per-capita use rise at a declining rate (Rock, 2000). If this change to more efficient water use tends to occur as a country's income status rises from the upper-middle to the upper income category, it could explain why per-capita water use seems to level off. The water-use intensity increase (or efficiency drop) in earlier structural changes from agriculture to industry and services may also explain the sharp increase of per-capita water use from the lower-middle to upper-middle country income categories.

Within countries, too, income level is important in determining the structure of freshwater demand. The poorest people in a country tend to use less water than those with high incomes. In cities, wealthier individuals are more likely to be connected to piped water supplies and waterborne sanitation systems, as well as to consume larger amounts of products with high embodied water intensities (the amount of water needed throughout the whole economy in order to provide final consumers with one dollar's worth of various goods or services, or in other words, the amount of water embodied in that one dollar's worth of quantity). Therefore, these individuals are more likely to be more profligate water users (Lenzen, 2002). Rural dwellers, who in developing countries are more likely to be poor, use less water because of inaccessibility of supply and lower consumption of goods and services.

Climate and culture can overcome income effects. On the other hand, a higher share of irrigated cropland or a larger natural endowment of water available to a country implies greater water-use intensity, over and above income effects (Rock, 2000). Because many lower-income countries are dependent upon agriculture as their primary industry and employer, they often devote a large part of their water supplies to agriculture. Developing countries with high levels of irrigation, such as India, are especially likely to use a large proportion of their water for agriculture. On the Indian subcontinent, 92 percent of water withdrawals are distributed to agricultural uses (FAO, 2003a). Lower-efficiency methods of irrigation used on over 90 percent of all irrigated land worldwide often use twice as much water per irrigated area as high-efficiency irrigation techniques such as drip irrigation (Postel, 1997). In low-income countries, about 87 percent of freshwater withdrawals go toward agriculture, on average, as opposed to 75 percent in middle-income countries and 30 percent in high-

Figure 4.8
Water Withdrawals for Agriculture (Percent of Total Withdrawals)

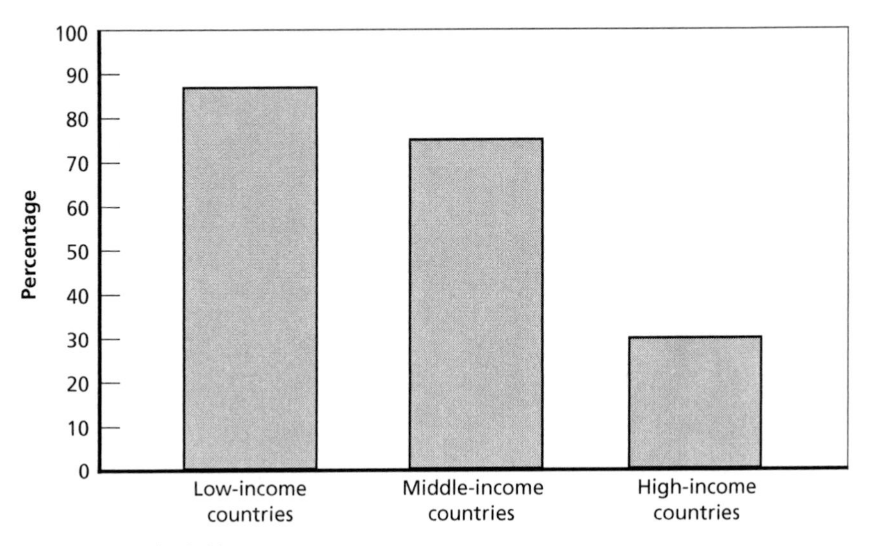

SOURCE: FAO (2003b).
RAND *MG358-4.8*

income countries (Figure 4.8) (FAO, 2003b). In general, the higher the level of development, the higher the proportion of water used for domestic and industrial purposes and the lower the proportion used for agriculture.

There are some important exceptions to this rule, however. In arid and semi-arid regions, both rich and poor, irrigated agriculture uses 73 percent of water withdrawn (El-Ashry, 1995). In many of these regions, crops must be irrigated most of the year. If no exogenous water (from outside country borders) is available, these arid countries have little chance of being self-sufficient in agriculture (Falkenmark and Widstrand, 1992). Additionally, in some rain-fed lower-income countries, relatively small proportions of freshwater withdrawals are used for the agricultural sector. Cameroon, which devotes 35 percent of its total freshwater withdrawals to agriculture, 19 percent to industry, and 46 percent to domestic uses, is one example; Congo and Nicaragua are others. Cultural influences can create exceptions as well. Japan, for example, still uses the largest share of its freshwater for irrigating rice, for reasons of taste preference, food security, and as a symbol of independence. As a matter of tradition, the Japanese government strictly prohibits the import of rice as long as local crops provide enough for the country (Krock, 2000).

CHAPTER FIVE

Influences of Water Resources on Demographic Variables

While demographic variables and their influence on water resources are a popular topic in papers discussing world water resources, especially by advocacy groups, the reciprocal influence of water resources on demographic variables is far less discussed in the context of water resource issues. However, the results of such influences are equally important, and are directly or indirectly responsible for millions of cases of illness and death worldwide every year.

Urban areas often create waste in quantities greater than the amount that can be absorbed by the surrounding environment. Growing cities often devalue and reduce their own water sources via pollution, land-use changes, and overextraction (IUCN, 2000). At the same time, they mobilize new sources farther away, extending the city's ecological footprint while raising costs of water provision to higher and higher levels. This increased pressure on land is created by the land requirements for industry, transport, and recreational activities that accompany the expansion of urban areas.

Land use of various types impacts groundwater and surface water quality. Groundwater is generally vulnerable to anthropogenic pollution from urban, industrial, and agricultural activities; naturally occurring contamination; and wellhead contamination. Industrial and some commercial developments pose the greatest risk, because of accidental and intentional spills of hazardous chemicals during transport, storage, and use (Water and Rivers Commission, undated). Also of special concern are horticulture, intensive animal industries, and urban runoff and leakage from septic tanks and other underground storage tanks.

Access to water and sanitation systems is a key health issue, as those who have less access tend also to have higher rates of disease, both because of poorer drinking water quality and because of reduced availability of water for hand washing (Giles and Brown, 1997). Countries with higher national incomes have built sewer networks and wastewater treatment facilities that have reduced the incidence of waterborne diseases. In contrast, rapidly expanding cities in much of the developing world have not been able to build such an infrastructure, and have high rates of illness from such diseases. Lack of access can be caused either by lack of infrastructure or by scarcity. In both cases, the results are similar: increased mortality and morbidity, as well as increased workloads for women and children, in particular, who usually shoulder much of the responsibility for water provision.

Waterborne diseases are spread when drinking water is contaminated with pathogens from the excreta of infected humans or animals and then ingested by uninfected humans. Hygienic and safe drinking water and sanitation are effective tools to combat waterborne diseases. One study in Tanzania found that women with a conveniently close supply of water had five percent lower child mortality than those who had to go farther, while women using an improved, protected water source had lower child mortality than those using a traditional open source (Evison, 1996, in Zaba and Madulu, 1998). A study in Malaysia found that safe drinking water and sanitation decreased infant mortality, especially among infants who were not breastfed or who were breastfed for a limited duration, since these infants are more likely to be fed formula mixed with contaminated water (see, e.g., Butz, Habicht, and DaVanzo, 1984). Many other studies have echoed these results (see, e.g. Shi, 2000; Feachem, 1981; Merrick, 1985; Patel, 1981; Schultz, 1979; WHO, 1978; Puffer and Serrano, 1973).

Globally in 1999, 82 percent of the population had access to "improved" water supply, meaning piped water, a public standpipe, a protected well or spring, or collected rainwater. The percentages are higher in developed regions and lower in less-developed areas (Figure 5.1). Improved sanitation (connection to a sewer or septic tank or access to an improved latrine) was available to 60 percent of the world's population in 1999, ranging from 48 percent in Asia to 100 percent in North America (Figure 5.2). It is notable that Asia is less served, on a percentage basis, than is Africa, despite the economic differences between the two regions, and despite the fact that Asia is better served in improved water supply. Although coverage is still far from universal, especially in sanitation, great progress has been made over the past 30 years, with large absolute increases in access to water and sanitation accompanied by substantial percentage increases, despite high population growth rates.

Figure 5.1
Improved Water Supply Coverage by Region, 2000

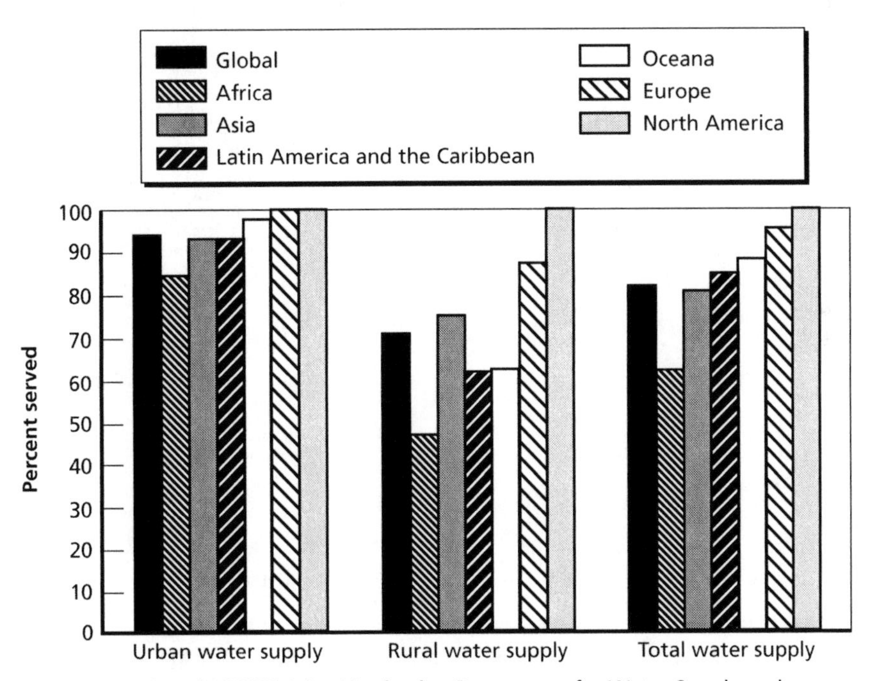

SOURCE: WHO and UNICEF Joint Monitoring Programme for Water Supply and Sanitation (2000).
RAND *MG358-5.1*

Figure 5.2
Improved Sanitation Coverage by Region, 2000

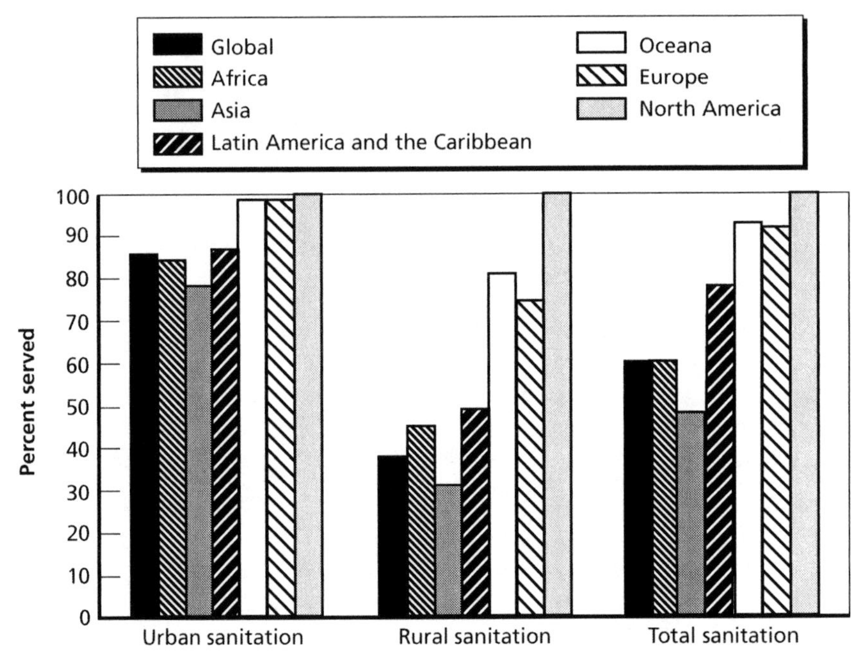

SOURCE: WHO and UNICEF Joint Monitoring Programme for Water Supply and Sanitation (2000).
RAND *MG358-5.2*

Migration

Water resources can affect migration decisions. Lack of water can be a factor in the decision to migrate away from a location. Deforestation, desertification, drought, and lack of land to cultivate can significantly influence migration from rural areas to towns (Falkenmark, 1994). This cycle can be caused or exacerbated by population growth. About 25 million people are estimated to be environmental refugees as a direct result of river pollution (de Villiers, 2000). Many others have become refugees due to the destruction of environmental resources that they require to live or to make a living. In one case, since 1975, India has diverted the dry-season flow from the River Ganges at the Farakka Barrage before the water reaches the southwestern part of Bangladesh. This has encouraged some people there to migrate to the bordering Assam region of India because of the environmental destruction caused by this diversion and because they do not have enough water to irrigate their crops. The migration has led to ethnic conflict in the Indian villages that received the migrants (Swain, 1996). Indian water and environmental policies in this case have led to the loss of livelihood for the poorest of the populace of the affected area in Bangladesh, perhaps two million people.

Many cases of forced migration due to the effects of dam building have been recorded. When a large dam is built, land upriver from the dam is flooded, causing the migration of any people living there. Dams have physically displaced millions of people worldwide since 1960 (Table 5.1) (WCD, 2000). In China and India alone, between 20 and 30 million people were moved to make way for dams in the second half of the 20th century (Cernea, 1988). The planned construction of large dams in India, China, Zimbabwe, and Brazil have

Table 5.1
People Forced to Migrate Due to Dams

Project	Country	Completed	Resettled People
Sanmenxia	China	1960	870,000
Akasombo	Ghana	1965	80,000
Aswan	Egypt	1970	120,000
Maduru Oya	Sri Lanka	1983	200,000
Itaparica	Brazil	1988	50,000
Ataturk	Turkey	1991	40,000
Yantan	China	1992	40,000
Yacyretâ	Argentina and Paraguay	Operational 1999; construction not complete	50,000
Sardar Sarovar	India	Under construction	250,000
Three Gorges	China	Under construction	1,300,000

SOURCES: World Bank (1994), in FIVAS (1994); Gleick (1993); Cernea (1988).

stirred intense debate and protests within those countries and internationally. The Three Gorges Dam in China, which will relocate about 1.3 million people, and the Indian Sardar Sarovar Dam, expected to displace over 250,000, are examples of controversial large dams under construction (Thackur, 1998, Cheng, 2002).

Others have migrated as a result of incremental deterioration of the environment. This kind of migration is most likely to affect the poor, who often live in marginal ecosystems and do not have the economic flexibility required to engage in long-term conservation measures at the expense of immediate environmental returns (Bates, 2002). Migration stemming from environmental deterioration and resource depletion is expected to increase in the future (Myers, 1993).

Freshwater availability can be a push factor for migration in developing countries, but is only rarely a pull factor, except perhaps in short-distance intervillage migration and movement between neighborhoods in urban areas (IUCN, 2000). This is partly because many rural-to-urban or urban-to-urban migrants settle on the outskirts of towns where no sanitation or water services are provided. When this happens, many migrants are forced to procure water either from distant and generally unclean sources, or to pay for water provided by private informal sector vendors, a decision that can cost them up to 40 percent of their income (Baer, 1996). In early societies, water was one of the primary factors for settlement, and there is still some evidence of this. For example, in the 1980s in the United States, in the Oglalla aquifer region of Kansas, an association was found between groundwater use and population change, implying active responses to access to groundwater (White, 1994). More commonly, however, water can be an indirect pull factor, when availability of water and other infrastructure induces industry to locate in a region. The employment opportunities created thereby may influence the scale of future labor migration (Zaba and Madulu, 1998).

CHAPTER SIX

Approaches to Sustainable Water Management

As the previous chapters have made clear, various demographic factors are working together to place stress on water resources by increasing demand for water, while decreasing supplies through pollution and destruction of freshwater ecosystems. At the same time, polluted waters are adversely impacting human populations through disease and mortality. Whether or not a water crisis is imminent, measures need to be taken to reduce the pressures on water resources well in advance of their collapse. Measures aimed at ameliorating this seemingly intractable situation can approach the challenge in different ways. One is to attack the problem by influencing the demographic factors themselves—working to reduce population growth, change population composition trends, and so forth. There has been progress on this front, and the recognition of the factors that are most likely to impact or be impacted by water resources will allow planners to track these factors and, perhaps, influence them in a way that will increase water supplies or lessen water demand. This is a subject that has been covered in great detail elsewhere, and is outside the scope of this monograph.

Another tactic is to reduce the impact of these demographic factors. This can best be done through management practices. The management of water supply and demand can make large differences in water withdrawals, and can influence the quality of water and its impact on human health. Two locations with similar demographic and ecological profiles may have very different water demand and supply profiles due to any combination of the following:

- differences in management approaches
- factors both within and outside of planners' purviews such as institutional arrangements, government subsidies, land tenure, government corruption, and political factors
- technological innovation, market forces, globalization of production and trade, poverty, and armed conflict.

Supply management involves the location, development, and exploitation of new sources of water. Demand management involves the reduction of water use through incentives and mechanisms to promote conservation and efficiency. Because the distinction between the two is not always clear, supply management can be defined as actions and policies that affect the quantity and quality of water as it enters the distribution system, while demand management governs actions and policies affecting the use or wastage of water after that point (Rosegrant, 1997). In order to meet the needs of water users and the environment into the future, both supply and demand management approaches will need to be utilized effectively. The level of each will vary with the amount of actual and perceived water scarcity and level of development in an economy (Rosegrant, 1997). Demand management increases significantly in importance as populations and economies grow and the competition for and value of water increase. The benefits of efficient allocation of water become ever more important as these factors increase, generally as a result of a maturing economy, but also as the result of physical water scarcity.

Supply Management

Options for increasing water supplies include building new dams and water-control structures (temporal reallocation), watershed rehabilitation, interbasin transfers (spatial reallocation), desalination, water harvesting, water reclamation and reuse, and pollution control (Meinzen-Dick and Appasamy, 2002; Winpenny, undated).

Dams and Water-Control Structures

The costs of developing new water sources, and particularly of large-scale supply solutions, have become prohibitively expensive, and have reached their financial, legal, and environmental limits in most industrialized and some developing countries (Frederick, Hanson, and VandenBerg, 1996). Lending by international donors for irrigation projects has declined significantly since its peak in the late 1970s, as have total public expenditures for irrigation in many countries, especially in the developing world. The growth of irrigated land area has declined in recent years, after nearly doubling in the first half of the 20th century, before more than doubling again between 1950 and 1990 (PAI, 1993). This is due partly to investment increasingly having to be made in rehabilitating existing systems, but also because development costs are so high (Pretty, 1995). The decline has been greater in developing countries, where investment in irrigation has been lower for a longer period.

Dams and water-control structures are often extremely controversial. Currently, there are more than 41,000 large dams (defined as dams over 15 meters high) in the world, impounding 14 percent of the world's annual runoff (L'vovich and White, 1991). The dams are used for energy production, flood control, and water storage, as a way to reallocate water temporally. Most of the best dam sites in the developed world are already taken, and dam construction there has all but stopped. In the developing countries, however, demand and potential is still high (WRI, 2000). As of 2003, there were 1,500 dams over 60 meters high under construction, with the largest number in Turkey, China, Japan, Iraq, Iran, Greece, Romania, and Spain, as well as the countries of the Paraná basin in South America (Brazil, Argentina, Paraguay, Bolivia, and Uruguay) (WWF, 2004). Huge benefits may accrue from large dams, such as electric power production, irrigation, and domestic water provision. However, dams have equally large environmental and social consequences, such as the submergence of forests and wildlife, the contribution to global climate change of greenhouse gases from the decaying of this submerged vegetation, the impact of changed flow in the dammed river, the relocation of thousands of families (as discussed in the section on migration) and the destruction of villages and historic and cultural sites, and the risk of catastrophic failure, among others. Additionally, there is a significant consumptive use of water associated with dams, as the evaporative losses can be large, depending on climate and the surface area of the reservoir. In Egypt, for example, a hot and arid climate results in annual evaporative losses estimated at nearly three meters (or 11 percent of reservoir capacity) at the Aswan High Dam. This compares to evaporative losses ranging from 0.5 meters to 1.0 meter in various parts of the United States (Rosegrant, 1997).

The planned construction of large dams in India, China, Zimbabwe, and Brazil has stirred intense debate and protests within those countries and internationally. One reason for the public reluctance toward large dam projects is that the social and economic costs of construction are borne by a relatively small group, while the benefits are widely distributed. It is possible that the resistance to such projects would be lessened, though not eliminated, if those bearing the costs of the project were adequately identified and equitably compensated. Additionally, the costs of not proceeding with a project are often not analyzed. A careful, impartial, and comprehensive cost-benefit assessment may help to clarify the alternative futures associated with the construction of a large-scale irrigation or other dam project, as well as the alternative solutions to the problems being addressed.

One dam, the Salto Caxias in Brazil, took an approach to dam building that created less controversy. The builders of the dam engaged the people who lived in the area that would be inundated and worked with them to establish resettlement priorities, and then added environmental priorities such as new farming methods, tree planting, and the establishment of a national park on the land surrounding the new dam. The cooperation with displaced families eased many of the social problems associated with dam building, but concern remains over the environmental implications of the dam, despite the efforts made to relieve them (Sutherland, 2003).

Watershed Rehabilitation

As discussed in the previous chapter, one of the best and most cost-efficient ways to secure water supplies around cities is conserving forests (Dudley and Stolton, 2003). As discussed, when watersheds are deforested, rainfall turns into runoff, which causes erosion, and sediment transfers and leaves no possibility of water storage, unless a dam is used. In areas of heavy seasonal rainfall, as in many tropical areas, especially in many developing countries, this effect is exacerbated. In developed countries, watersheds are often protected, maintaining vegetative cover and limiting or eliminating human, livestock, and agricultural use of the land. In most developing countries, governance of watershed use is weak, and the long-term benefits of nonuse or rehabilitation fare poorly against short-term political or human goals (McIntosh, 2003). This is especially true in densely populated poorer countries, where competition for land and resources is fierce. It is argued that reforestation of degraded and unproductive land is cost effective in the long term if all the ensuing benefits (e.g., water supply, agricultural revenue, pollution reduction, and fuel and timber benefits) are taken into consideration ("Dry Future," 2000). The high initial cost combined with the incompatibility of long-term planning with short-term political thinking, however, make watershed rehabilitation a difficult decision in many places.

Small-Scale Irrigation Systems

Given the problems surrounding large-scale water projects, small-scale irrigation systems may provide an opportunity to expand irrigation with fewer constraints. Studies in sub-Saharan Africa have shown that small-scale irrigation systems are not inherently more or less successful than large-scale irrigation projects. Instead, their success depends on institutional, physical, and technical factors. Successful small-scale irrigation systems there had the following characteristics:

- simple, low-cost technology (usually small pumps)
- private and individual arrangements for operating the system
- supporting infrastructure adequate to allow the sale of surplus production
- high and timely cash returns generated for farmers
- farmers who are active and committed participants in the project design and implementation (Brown and Nooter, 1992).

Because each watershed is different, the large-versus-small distinction is only one of many variables that must be taken into account when projects are being considered. Both large and small projects must involve broad accounting of the project costs and benefits, including environmental and social externalities and benefits in all sectors, and fair compensation for negative impacts. Such decisions should be taken in consultation with those who will be impacted by the project, both positively and negatively.

Groundwater

The careful exploitation of groundwater resources is another potential option for increasing freshwater availability in many countries. Several large aquifers lie under the arid Middle East and North Africa regions, for instance. These aquifers store volumes of water equal to several years of the annual average renewable supply of the region, but recharge rates are often 2.5 percent of total volume stored per year or lower, which means that sustainable use of the resource can tap only that rechargeable quantity per year. Much of Asia and parts of Latin America have untapped groundwater potential. In much of sub-Saharan Africa, aquifers are small and discontinuous, with slow recharge rates, so use will be limited to local and regional areas.

Groundwater is generally a more efficient source of irrigation water than open canals. This is because the water can be accessed near to where it will be applied, reducing transportation losses; in addition, farmers can control the amount and timing of the water, giving an advantage over irrigation from surface water. In India, crop yields from farms irrigated by groundwater were found to be 1.2 to three times greater than farms irrigated with surface water (Shah, 2000).

Groundwater utilization has been seen as a means to alleviate poverty, since anyone who can afford to install a pump has access to free water (Moench, 2002). The problem is that, as with any common resource, the incentive exists to waste and overuse the resource. This has led to overuse of many aquifers and the falling of many water tables. Also, since groundwater development is undertaken, for the most part, on an individual basis, there is much less oversight and a much smaller body of law governing the allocation, management, or monitoring of groundwater (FAO, 2003b). This and the fact that groundwater, unlike surface water, cannot be readily seen, make conservation measures and improved efficiency of use of groundwater extremely difficult. Also, in developing countries, groundwater storage and recharge rates are generally poorly understood. An increase in the knowledge of aquifers and their characteristics would help to maintain sustainable use of this resource.

Interbasin Transfers and Exports

Interbasin transfers are another way to provide water to areas with dense populations, such as urban areas, or other water-short areas. Canals and pipes are the usual means of transporting the water, though new technologies such as large water bags attached to ships are another option that has been used for freshwater deliveries to emergency situations and for coastal areas with a need and desire for expensive water (Gleick, 2001b). Such large infrastructure projects are expensive and potentially environmentally harmful, both to the freshwater ecosystem from which the water is extracted and the lands over which the pipes or canals must flow. Additionally, inhabitants of the land lying in the path of the pipes or canals are demanding equivalent access to the water, adding to the demand from the source (Meinzen-Dick and Appasamy, 2002).

Water exports, where water is sold across state or national borders, are a controversial subject. Although the spatial reallocation of water makes theoretical sense, the selling of water, more than the sales of other resources, seems to strike a nerve in the public. Two recent plans to sell water from water surplus areas have met with resistance. The controversy in both cases was the idea of corporate profits taking precedence over environmental concerns, and the issue of water as a common resource held in public trust. In Michigan, a private company wanted to bottle spring water for sale within the United States, and in Canada, a private firm wanted to transport water in tanks to sell to drought-stricken U.S. states, (Schneider, 2002; Edmonds, 1998). Residents in both cases felt that the water should not be given to a private firm to sell for private gain with no benefits, and perhaps some costs, such as environmental repercussions and resource depletion, falling to residents. There is also concern surrounding the ways in which private companies get access to water supplies. Though further discussion is beyond the scope of this paper, it is important to note that this sort of public opposition to the export of water resources comes up repeatedly. Barriers of this kind will pose a challenge to the use of this option for the augmentation of water supply.

Water Reallocation

The competition between cities, industry, agriculture, and environment for water resources is an issue that is expected to become more contentious over the next several years. Even though total domestic water use is far less than agricultural use, the rapid growth of urban areas will result in the growth of water demand in some locations. One of the ways of supplying increased demand in both the urban domestic and industrial sectors will almost certainly be a reallocation of water to these sectors from the agricultural sector.

This is already happening in many places. In the western United States, agricultural interests that had nearly unlimited access to cheap water are finding their access restricted and their payments increased. In the developing world, reallocation is also occurring, though often in a more informal manner. In India, for example, well owners and irrigators pump water from rivers into tankers, who then sell the water to households in the urban areas (McKenzie and Ray, 2004). This informal market, though illegal, is possible because of the differential value of water across sectors and because of lack of infrastructure and access to water in the domestic sector. Wealthier households connected to the public water system pay a subsidized rate more than ten times less for water than lower-income households buying from informal water sellers, and farmers can grow crops with low water requirements, sell their remaining allotment of water, and make more than their colleagues who grow traditional crops utilizing their entire water allocation.

Water reallocation can be done via supply management, by reallocating water from the top down, or via demand management, using incentives to move water between sectors. Because water in developing countries is predominantly consumed by the agriculture sector, reallocation of even small percentages of water consumption to the domestic sector can fulfill domestic needs.

Although there are many examples of farmers willingly sharing their water supplies when compensation is offered, most notably in the western United States, Mexico, and Chile (ECSP, 2002), there is a limit to the reallocation of water away from agriculture, as food production has a strong political component because of rural jobs, food as a basic need, and food security. A reduction in economic activity associated with the reallocation of water away from the agricultural sector could limit future economic development in a region and even induce out-migration (Rosegrant, 1997). However, in Chile and California, the effects have been small to this point, with agriculture revenues dropping only two to three percent in California counties where water was sold (Dixon, Moore, and Schechter, 1993). In Chile, farmers sold only small portions of their water allocations and practiced very efficient irrigation with their remaining water, resulting in only small revenue reductions (Rosegrant, 1997). This highlights a positive outcome of water reallocation: an increase in agricultural conservation practices, which become more economically appealing with high water prices. That is not to say that economic effects might not be large if participation were greater and farmers began selling larger portions of their allocations.

Desalination

Desalination is the process of turning saltwater into fresh drinking water through the extraction of salts. Desalination through evaporation (or distillation), in its simplest form by the sun, has been used for many years, but large-scale desalination by evaporation is an energy-intensive process that is extremely expensive. Additional costs are incurred when the water is moved from the sea to its destination. Currently, mostly wealthy, energy-rich, water-scarce countries, such as Kuwait and Saudi Arabia, and Israel use desalination on a large scale, although Tampa, Florida, and Houston, Texas, among other cities in the United States, are constructing desalination plants to reduce pressure on groundwater in those areas. Island nations such as Cyprus, Trinidad, and the peninsular city-state Singapore have also invested in desalination plants. Desalination accounts for only 0.2 percent of water withdrawals worldwide (Martindale, 2001). Currently, desalinated water is delivered to users at a cost of $1–4 per cubic meter, comparable to the costs of new water supplies in the most arid areas of the world, but very high compared with alternative sources in much of the rest of the world. For example, desalination costs are well above the rates paid by urban users in the United States of about $0.30 to $0.80 per cubic meter (Gleick, 2000a). Transportation costs are high for water, so attempts to use desalinated water in inland areas incur a significant cost penalty. A desalination method called membrane desalination uses reverse osmosis to produce freshwater with less energy, and therefore more cheaply, than evaporative methods. This is the method being used in the Tampa and Houston plants, and it is possible that this method will prove cost effective over a wider range of applications.

There are other issues associated with desalination, especially the disposal of the brine stream produced during the process. If desalination occurs near an ocean, with care the stream can be disposed of there, though there are reports of marine pollution from the dumping of high-temperature brine (Barlow and Clarke, 2002). Also, the energy used to fuel the desalination process is itself generally polluting, and even may use water as one of its inputs. Although solutions such as the use of renewable energy to fuel desalination plants may solve this problem, they would add unacceptable costs to the process.

Desalination research programs continue to make advances that will bring the price of desalination down. Although it is unlikely that desalination will make a large dent in the supply of world water in the near future, it is important in certain situations, such as for domestic and industrial purposes in coastal areas of water-scarce, relatively wealthy countries. Technology and costs may change, and desalination may become more important in time. At the moment, however, the expense, both economic and in terms of the environmental impacts of energy use and brine disposal, seems high, and this supply option, like many technological solutions, may be better left as a last resort option.

Water Harvesting

Water harvesting can be as simple as a water barrel under a rooftop gutter, or as complex as a sub-watershed scale storage and groundwater-recharging scheme. The capture and diversion of rain or floodwater to fields for irrigation is an old tradition in many agricultural areas, and the traditional technology can be more widely used and improved upon to increase production and income on farms in some areas. Traditional techniques include stone or earthen berms constructed across fields to either contain or redistribute rain and floodwaters, and vegetative barriers that slow runoff. Often these techniques also reduce irrigation in fields. In the urban areas in the domestic sector, water harvesting, now required in new buildings in some Indian cities, can provide positive environmental benefits by reducing the polluted runoff from urban areas into surface waters, along with reducing water use from other sources.

The storage of runoff during the rainy season is particularly important in climates where much of the available rainfall falls within a short period. The water collected can be used either directly or to recharge the groundwater, and its generally small-scale and environmentally friendly nature makes water harvesting appealing (Meinzen-Dick and Appasamy, 2002). The water that is harvested would have otherwise run off into, for example, seasonal rivers or oceans; water harvesting, by changing the temporal distribution of water, may modify the natural cycle in either positive or negative ways, particularly in rural areas. Water harvesting is a good example of more efficient use of green water, and can be an important way to increase efficiency, productivity, and soil fertility on a local and regional basis.

Water Reclamation and Reuse

More important in the industrial and domestic sectors might be reuse and recycling of water. Pure water is not needed for many tasks such as irrigation, and some types of washing and cooling tasks, as well as environmental and ecosystem restoration and the matching of water quality and water requirements can be an important part of better water management. In developed countries, most notably Japan and the United States, industrial recycling of water has been an important component of water-use reduction in the industrial sector, and in developing countries there is a movement toward seeing wastewater as a resource, rather than a problem. It is a cost-effective and reliable water resource despite significant social and economic barriers to wastewater reuse in some regions (IUCN, 2000). The advantages of wastewater reuse include water-pollution abatement, not discharging into receiving waters, and providing long-term water-supply reliability within the community by substituting for freshwater.

Water reclamation and reuse occurs when water is used, recaptured, and cleaned for reuse. Nonpotable reuse of water for irrigation of agricultural fields and landscapes, industrial cooling, and some commercial indoor uses are the most commonly encountered. The direct potable reuse of wastewater is not popular with the public, and its use is limited to date to only Windhoek, Namibia, and Denver, Colorado, in the United States (Shaw, 1999). In the future, there may be more scope for direct reuse, but the relatively high cost of treatment and community attitudes make it likely that other uses will be expanded first.

Only a small portion of industrial water, for example, is actually consumed. In most cases, water is withdrawn and returned to the freshwater system at a lower water quality. Instead, the remaining water can be treated to any level of quality, at different costs, or may be reused within a factory without treatment. This results in the reduction of both water requirements and effluents, improving the output per cubic meter of water. In developed countries, water recycling in industry has become commonplace, and has resulted in a reduction in the total industrial water use as well as the water use per dollar of output in many countries, especially Japan and the United States. Between 1973 and 1989 in Japan, for example, total industrial water use declined by a quarter, while industrial output per cubic meter of water used declined from $77 to $21 over that same period (Postel, 1997), and has remained stable since that time (Maita, 2003). In 1998, the ratio of recycled water to industrial water use was 80 to 90 percent in the steel, chemical, and allied product industries, compared to 20 to 40 percent in the paper and allied product, plastic product, and textile industries, which may be used as a goal for similar industries in other countries (Maita, 2003).

Because wastewater is discharged year-round, treated wastewater is a good match for industrial use. In Madras, India, several large industries have been paired with the local water agency to use partly treated sewage (Meizen-Dick and Appasamy, 2002). The industries purchase and treat the sewage, then use it for cooling and manufacturing processes. Both the industries and the water agency gain from the arrangement—the water agency gains revenue, reduces pollution, and saves treatment costs, while the industries have a reliable source of water, something that was not assured previously.

Wastewater and gray water, which are generally nutrient rich, can also be reused for food production and other tasks that do not require pure water. Grey water from showers and sinks, for example, can be used directly in gardens or stored in soak pits for groundwater recharge. Reclaimed water is also increasingly being used to augment natural groundwater recharge and for environmental and ecosystem restoration (Gleick, 2000a). Wastewater reuse measures are already in place in some arid areas such as Israel, Portugal, Tunisia, Namibia, and California, as well as in Japan. Countries such as Morocco, Jordan, Egypt, Malta, Cyprus, Greece, and Spain, as well as France and Italy are considering or actively pursuing wastewater reuse schemes. In Israel, for example, treated wastewater provides 18 percent of the country's total water supply, and nearly 40 percent of the agricultural water supply. The most successful wastewater-reuse irrigation projects have been located near cities, which can provide both wastewater and a market for agricultural products. The rate of expansion of wastewater reuse is dependent on both the final quality of the wastewater and the willingness of the public to embrace the use of wastewater supplies (Rosegrant, 1997).

The leading problems associated with wastewater reuse are high salinity levels and the accumulation of heavy metals in the water, which can make the water unusable in most applications, and can increase the salinity level in agricultural soils when used as irrigation water. There are also concerns about risks to workers and consumers from bacterial contamination, but this problem can be easily controlled with adequate treatment of water, if necessary, before reuse.

Pollution Control

A final supply option is to maintain the quality of the current supply of water. If water sources are protected from contamination from agricultural residues, soil erosion, runoff from urban areas, industrial effluent, chemicals, excess nutrients, algae, and other pollutants, the current supply of water can be retained, the cost of developing alternative supplies can be saved, and previous sources can be opened up after they are cleaned (Winpenny, undated).

Controlling pollution reduces the impact of water on populations by maintaining public health, maintains freshwater ecosystems and ensures their continued provision of goods and services, and provides recreational opportunities. Although most countries have pollution-control laws, including effluent standards that are required to be met by all polluting industries and municipalities, many, especially in developing countries, have not had the political will or financial resources to enforce them. Pollution control laws in developed countries have helped to clean up rivers, lakes, and streams and have promoted conservation and efficient use of water, but they have also promoted the more efficient use and conservation of water, since the most cost-effective way to meet pollution limits and water quality standards is often to reuse and recycle water, especially in industrial processes.

In many cases, municipalities may not be able to build sewage treatment plants with enough capacity to handle large and growing loads of effluents without financial assistance from higher levels of government. Even with assistance, industries will generally have to either pretreat their wastes or invest in individual or common treatment plants. However, because industries in developing countries have been reluctant or unable to do this, incentive measures are needed.

As in higher-income countries, effluent charges based on the quality and quantity of effluent discharged can encourage industries to conserve water and reduce pollution (Meinzen-Dick and Appasamy, 2002). Of course, this must be coupled with stricter regulations and enforcement, and strengthened by public disclosure and education.

Some unexploited water sources can still be harnessed, but few are left that would not have some impact on environmental systems, or that are readily accessible and economical. Therefore, the most economic option for water planners may increasingly be to meet additional requirements by reducing demand by any number of means, and by looking at alternatives such as water reclamation and water harvesting. Not only are these options often the most appealing from a strictly monetary standpoint, they are also often the most appealing from an environmental and public acceptance standpoint. Water users seek the services that water provides, and if the same services can be provided reliably in a way that causes less damage to the natural environment or at lower cost, water resources will be further protected for future use.

Demand Management

Although supply-management measures such as those discussed previously will help meet demand for water, much of the water to meet new demand will have to come through demand management, including conservation and comprehensive water policy reform.

Without appropriate management, no technical or engineering fixes will solve all of the problems of water supply and demand and the ecological damage that is associated with its misuse. Sustainable water development and management require the integration of social and economic concerns with environmental ones. Management approaches such as integrated water resources management and demand management offer effective means of providing water for human use while easing the stress on freshwater ecosystems and the ecological goods and services that they provide (CSD, 2001).

Overall, *poor governance* is at the core of many water problems, especially in developing countries. For example, relative water scarcity may be partly caused by too much water being allocated to too few activities with too high a subsidy (CSD, 2001). Poor governance results, in most cases, from corruption and outside interference, and is exacerbated by low tariffs, which lead to a host of problems such as low water supply and sanitation service, overexploitation of water resources, conflict between rural and urban users, and poor maintenance of existing water supply lines (McIntosh, 2003). Further mismanagement arises because the most efficient and effective management occurs at the scale of the watershed. However, the technical, economic, and institutional constraints on performing water management on that scale are formidable.

Recall that there are three types of water scarcity: absolute, economic, and induced. Usually scarcity in a particular location is caused by a combination of two or three of these scarcity types. *Demand management involves saving water from existing uses by reducing losses.* Losses can also be of any of three types: absolute, economic, or induced. Absolute losses occur, for example, when water vapor escapes to the atmosphere through evaporation or evapotranspiration, or when water flows to salt sinks such as oceans or saline aquifers. Economic losses occur when water flows to freshwater sinks, but is economically unavailable due to its location, when water is polluted to the extent that it is no longer economical to treat it to a usable quality, or when drainage water is too expensive to reuse due to factors such as physical characteristics of aquifers, deep percolation, slopes, and lifts (Rosegrant, 1997). Induced losses occur, for example, when the effective cost per unit of lightly polluted water is higher because crop yields per unit of water are lower (Rosegrant, 1997), or when low tariffs or high subsidies misallocate water. If the effective price of water and the true scarcity value of water are greatly different, water will be lost to the system. The effective price of water consists of more than just the tariff and subsidies; it also involves physical considerations and other factors.

The point of demand management is to reduce the losses in all of these categories. In order to do so, comprehensive policy reform may be required in many places. Because of entrenched interests, cultural and religious considerations, and tradition, policy reform, like all reforms, may be very difficult and slow. However, the reforms are generally needed if water scarcity is to be averted, since, as discussed earlier, supply augmentation alone will not provide a sustainable water supply indefinitely. Several categories of policy instruments can be used to enact demand management and improve governance of water systems. They include the following (Bhatia, Cestti, and Winpenny, 1995, in Rosegrant, 1997):

- *Enabling conditions:* actions that change the institutional and legal atmosphere in which water is supplied and used. This may include the broad improvement of governance by making policies transparent, the involvement of civil society, and regulatory bodies, as well as specific reforms such as water rights.
- *Market-based incentives:* actions that use economic incentives to directly modify the behavior of consumers. Policies include tariff reform, which may raise or lower prices for all or certain users, adjust subsidies and taxes, and include charges for pollution or effluent flows.
- *Nonmarket instruments:* policies that use direct, nonincentive-based laws and regulations to regulate water use. These include quotas, licenses, pollution controls, and restrictions.
- *Direct interventions:* technological and other interventions to conserve water. Examples include leak detection and repair programs, investment in improved infrastructure, and conservation programs.

The exact nature and combination of policies will vary from country to country, and even within countries. The choices made will depend on the existing quality of governance, institutional capacity, level of economic development, relative water scarcity, and other factors. Some examples of demand management policies that can make a difference in many countries follow.

Enabling Conditions

Water quality matching. As noted above, high-quality water is needed for human consumption and certain industrial processes, but many other uses can be met with lower-quality water such as storm water, gray (lightly used) water, or reclaimed wastewater that can be used for landscape irrigation and some industrial applications. In order to facilitate the reuse of water, single-pipe distribution systems can be replaced with integrated systems that can deliver different qualities of water to and from the places it is needed. These systems are cost effective and practical, especially in new construction and areas without existing piped supplies.

Management of water must meet the needs of people, rather than merely supply water. In other words, the emphasis should be on providing goods and services, not water per se. As a result, efficiency is part and parcel of the management approach, helping to reduce the impact of demographic factors such as population growth, decreases in household size, and rising standards of living, by ensuring that the least amount of water is used to provide the most services. Also, as discussed previously, it is important to distinguish between water needs and water wants, and reducing nonessential wants and wasteful practices when necessary. Inefficient water-resource utilization occurs when water is allocated to activities that are not in line with the socioeconomic and environmental objectives of a country or region, especially if other activities that would better serve development objectives are deprived of water. Most societies give direct human consumption of water highest priority, an objective that is codified in national water policies, and many also give priority to agricultural use (Meinzen-Dick and Appasamy, 2002). Conflict among users results, made worse when water distribution is too inflexible to adapt to changing supply and demand conditions (Frederick, Hanson, and VandenBerg, 1996). The institutional structure of water management must be revised in many countries to meet the needs of new economic and social imperatives, and to incorporate new information on ecological and human needs.

Decentralized supplies and involvement of users. Although not appropriate in every case, decentralized water systems are cost-effective once the very cheapest centralized solutions are exhausted, and reliable when users are involved in the planning, execution, operation, and maintenance of water systems (Wolff and Gleick, 2002). In peri-urban and suburban communities, decentralized water systems may be more economical than extending the centralized system and those communities may also prefer to manage their own systems (Meinzen-Dick and Appasamy, 2002).

In addition, poor communities in urban areas may not receive wastewater and water services at the same level as do richer communities. This is especially acute where formal and informal settlements exist in an urban area, as is the case in many cities in developing countries, but exists as well in different forms in urban areas of the developed world. Low-income communities may wish to have their own community-managed systems if they will improve performance or lower costs, as might be the case in an area without infrastructure where water is brought in by tanker trucks or kiosks (Meinzen-Dick and Appasamy, 2002).

A good example of the decentralized approach to groundwater management is in Southern California, where the governance structure is agreed upon and managed by water users, is responsive to local conditions, adapts to the dynamic environment, and uses existing data. It is, in many respects, very successful and efficient, and demonstrates the importance of water managers engaging communities and individuals in water management, and interacting closely with them to aid them in decision making for water systems.

Privatization. Some see the development of water markets and the privatization of water management as ways to allocate water more efficiently, although both methods are extremely controversial. Water markets can encourage conservation by allowing the trading or selling of water rights, for example. Private-sector participation in water-resources management and financing ranges from building to operating and owning municipal water systems. Because private firms operate under profit motivations, concerns about equitable access to water for the poor, the integration of environmental concerns, and the sharing of risks must be addressed before privatization becomes a viable option (UNEP, 2002).

There have been both successes and failures in the privatization of water systems. In general, however, it is understood that some private-sector investment is required in many places, especially in developing countries, if water systems are to be successful in the future.

However, the lessons learned by privatization occurring in the last several years should be taken into account, such as these (McIntosh, 2003; WSTB, 2002):

- *There are many forms of privatization of water services, and the form chosen must be adapted to the culture, political structure, and legal and regulatory framework of a particular location.* In many developing countries, there is resistance to the idea of international companies owning water systems. In these cases, a domestic owner or partnership may work better, as may a public-private partnership. An even playing field for all possible owners and operators should be a priority.
- *Improved performance by public water utilities has resulted in some cases due to pressure from large national and global water companies.* Support should be given to governments to help them meet water needs without private-sector participation, if viable, as a first step. Often, tariff reform and regulatory changes are needed elements of water sector reform whether the owners and operators of the water system are public or private, and true reform will be impossible without these elements. For example, in the United States, public urban utilities can now enter into contracts of up to 20 years for the operation of private participation in the operation of plants that were funded by public bonds or grants. This was made possible by the loosening of federal tax laws. Additionally, the Safe Drinking Water Act of 1974 has pressured small- to medium-sized utilities to improve their operations or seek assistance in the private sector.
- *Not all privatization efforts are successful.* In many cases, especially in developing countries, promises of efficiency and total income from water are unfulfilled. For this reason, the municipality must retain the ability to monitor performance and assume operations in case private operations fail. Regulatory arrangements must exist before privatization of assets, and must give the power to enforce such regulations to independent entities. Openness and transparency are equally important.

- *Good relations are necessary between governments and private operators.* Social, ecological, and cultural matters must be considered and agreed upon in consultation with governmental and local input.
- *Tariff revision, structures, and mechanisms for tariff setting must be agreed upon.* The public has shown its willingness to pay for reliable, high-quality water. However, many governments are reluctant to raise tariffs. This is essential to the success of water systems, and must be agreed upon in advance. The poorest sector of the urban population must be adequately served, often a major constraint in developing countries.
- *The water services industry faces a great need for maintenance and replacement.* This is true in both high and low-income countries, as demographic expansion has outpaced municipal investments. However, using privatization as a means of financing investment is unlikely to be the cheapest way. Host countries bear almost all of the risk of price demand or exchange rate, and investors assume almost none. Additionally, unlike debt, such a transaction does not attract relief measures (Lobina and Hall, 2003).
- *Some types of contracts have proven successful; others have proven risky to the host country.* For example, the design-build-operate contractual arrangement and its variants have been successful in some large water-service systems in the United States. However, take-or-pay agreements often oblige public authorities in developing countries to buy bulk quantities of water irrespective of future demand. In this case, public funds guarantee multinationals' profits at taxpayer and consumer expense (Lobina and Hall, 2003).
- *Good and reliable water sources are essential for reliability in the long term, and therefore will be critical to water services privatization.* This includes issues pertaining to the development of watershed lands and the protection of environmental quality around reservoirs. This is essential whether privatization of water services takes place, and is something upon which the local or national government involved must act.

- *Employee rights should be protected, including offering retraining or transfer.* Labor force issues are important both as a possible source of cost savings and a focal point for public concern.
- *Most water utilities will continue to be publicly owned and operated.* However, in the most troubled cases, private help will be needed, either permanently or temporarily. It is important that all contracts set up to transfer ownership or operation of water services to a private entity are carefully, independently, and transparently negotiated so that all parties wind up better off.

Economies of scope. Economies of scope can reduce the total cost of water systems by combining the decisionmaking process of several authorities, integrating thinking about land-use patterns, flood control, and water demands. Integrated water-resource management involves the management of water resources at watershed level. It recognizes the need for institutions and policies to consider the watershed basin as a whole when management decisions are made, to assure that, for example, interventions made upstream do not negatively affect users located downstream.

Several requirements must be met in order to realize the potential of integrated water-resource management. Adequate funding, human and institutional capacity-building, and a better grasp of the extent of the freshwater resource and supporting ecosystem need to be combined with education and training and the transfer of appropriate technological solutions. This is particularly important in areas in developing countries that are already water-scarce (UN, 2002a). As mentioned earlier, a critical step in improving water resources management is the recognition of the primacy of water catchment areas as the appropriate scale for effective water and ecosystem management. Management decisions taken at the level of the watershed allow all impacts to be accounted for, as opposed to decisions at the level of an individual water source. This is even more imperative in arid and semi-arid climates where changes in land use and vegetation have clear implications for land and water (SIWI, 2001). This will often require cooperation and coordination between countries or administrative units, which is extraordinarily complicated due to political prerogatives, but is nonetheless the best way for water resources and freshwater ecosystems to be managed competently and resources equitably distributed.

Market-Based Incentives

Water Pricing. The recognition of the social and economic value of water and of freshwater ecosystems allows the resources to be compared with other social and economic goods and reinforces their status as scarce and essential resources. Water supply has in many cases not been treated as a commercial enterprise. In most developing countries, farmers and households either do not pay for water per unit used, or pay a low price (Meinzen-Dick and Appasamy, 2002). Governments, businesses, farmers, and consumers must treat water not as a free good, as they often do now, but rather as a scarce resource that comes at a price. (Rosegrant, 1997). If water is seen as a free or nearly free good, the incentive for conserving it is absent. Limited access to water adds costs to water the same way that higher economic prices do. Water pricing can help to assure adequate supplies of water, as well as encourage environmentally appropriate usage. In the water-scarce arid and semi-arid western United States, farmers traditionally have paid nothing for water and only a small amount for its transport to their farms. They then have applied it liberally to low-value crops, imparting a marginal value of well under $50 an acre-foot—the amount of water needed to cover an acre of land with a foot of water (Frederick, 1998). The water would have more value left in stream to provide hydropower, fish and wildlife habitat, and recreation than it does when diverted to irrigation in this way. The value of water might also rise by selling it to urban areas that are spending ten times as much to augment supply in other ways (Frederick, 1998).

Keeping water prices low can be presented as controlling inflation or making service affordable to the poor. However, in reality, it ensures service deficiencies such that the poor never receive their supply and have to pay very high rates to vendors. Raising prices, on the other hand, can result in civil unrest, especially if it has to be done before service can be improved (Rogers, Kalbermatten, and Middleton, 1999). This may be prevented by raising prices gradually, and by involving the community in the implementation, especially if the stated goal of such price increases is to improve reliability and to invest in connecting the urban poor to piped water.

There are many benefits connected with increased prices, such as the following (McIntosh, 2003):

- Demand is reduced, conservation is made viable, substitutes are relatively cheaper, and consumption preferences change.
- Supply is increased, as marginal projects become viable and efficiency measures are economically attractive.
- Reallocation between sectors, especially from agriculture to domestic and industrial, is facilitated, as are reallocations between off-stream and in-stream uses.
- Increased revenues at utilities allow for improved managerial efficiency, partly by making modern monitoring and management techniques and staff training affordable.
- The per-unit cost of water to the poor is reduced if piped supply is extended, since many (most, in some places) poor people rely on expensive water from vendors.
- Environmental sustainability is more easily attained due to reduced pollution loads (especially due to recycled industrial water) and reduced demands on water resources, leaving more water available for ecosystems (Rogers, 2001).

A common problem in developing countries is the flawed nature of price structures. This is especially true when lifeline tariffs are set too high, as is frequently the case in developing countries. Often such tariffs benefit only those who have a house connection, who are more likely to be urban rich than urban poor or rural dwellers of any income bracket. Prices should increase with increasing consumption. Low-income users need to be protected by lifeline rates, but the higher levels of consumption should be charged at the marginal cost of developing new supplies (which is what excessive use will eventually force). This is likely to be two or three times higher than the cost of current supplies, and should act as a deterrent to frivolous use

(Rogers, Kalbermatten, and Middleton, 1999). Many studies have documented willingness to pay for water in developing countries with unreliable or nonexistent access to public water supplies. Many times, consumers are purchasing water from private vendors at a level (as much as 25 times the unit rate paid by the rich) that matches what would be required to finance the development of a water infrastructure, or are willing to pay large amounts relative to their income to secure a continuous supply of water (see, e.g., Whittington, Lauria, and Mu, 1991; Altaf et al., 1993; Brookshire and Whittington, 1993; McIntosh, 2003).

Nonmarket Instruments

Natural infrastructure. Natural infrastructures and their need for water must be included in management of water resources. Human users of water demand water-based services such as fishing, swimming, tourism, and the delivery of clean water to downstream users. Water that is not being withdrawn by humans goes to these and other productive uses. Reducing the amount and quality of water available for use by ignoring the importance of natural systems ignores the impact of water resources, especially their degradation, on health, mortality and other demographic factors. This has been well recognized in most developed countries, but has yet to become a factor of importance in most lower-income countries. Many demand-management instruments promote environmental sustainability and water quality, and the goals of use efficiency, and conservation, economic efficiency, and environmental sustainability are usually, when carefully examined, complementary (Rosegrant, 1997). California again provides an example of this: between 1960 and 1990, urban water use, water use in irrigated agriculture, and legally mandated runoff for environmental purposes all increased, despite competition among these sectors.

The challenge of water management is that, in order to guarantee adequate water for all users, there is a menu of trade-offs from which choices can be made. Some of these choices in allocation are between sectors and demands, in upstream-downstream water sharing, and in allocation of water between societal uses and ecosystems (SIWI, 2001). In the case of freshwater ecosystems, a management decision such as the building of a dam or the pumping of groundwater may do damage that is irreversible or widespread. Much of the damage that will be done may be unaccounted for by those managers making the management decisions. When humans manipulate soil, vegetation, and water systems, they produce benefits, but they also produce environmental feedback due to intricate interdependencies and interactions in the ecosystem (Falkenmark, 1994, p. 99). Unfortunately, the combination or degree of abuse that will bring about a system collapse is generally unknown, and the stress of a single small change may seem harmless. This makes it difficult to know how to alter management policies to ensure that ecosystems can be maintained at a level at which they are providing the full complement of goods and services that they originally furnished. All of this is complicated by the fact that the decisions made in the past by water planners were made under the assumption that future climatic conditions would be the same as past conditions. Because of the effects of climate change on climate and water resources, a reliance on the historical record may lead to incorrect and potentially disastrous decisions (Gleick, 2001b).

Education measures and civil society involvement. In order to maintain reductions in water consumption over time due to dramatic increases in water prices, measures such as education campaigns should be implemented concurrently (UNCHS, 1996). This is true because the conservation of water relies, ultimately, on behavioral changes. These changes are more likely to be achieved if the public is aware of and understands the reasons behind such changes. Education can take place through the media, schools, and user groups, and can inform the public about water conservation methods, sanitation, water scarcity, the need for water-tariff increases, if any, government policies and plans, and other topics (Meinzen-Dick and Appasamy, 2002). Civil society, which includes all stakeholders with interests in the water sector (e.g., consumers, nongovernmental organizations [NGOs], academics, journalists, and utilities), can become involved by putting pressure on the government and other water managers, as well as on the general public, to better manage the water resources, and ensure equitable distribution of resources. This can take the form of education or advocacy. Education can help communities understand the link between water, sanitation, health, and productivity, and help industries learn about the efficient use of water, efficient treatment and discharge, and the need for higher water prices, in some cases (McIntosh, 2003). It can also help to move toward transparency in the management of water systems, as educated consumers demand accountability from water managers and others.

Direct Intervention

Conservation. Demand management can include laws and regulations aimed toward conservation, such as restrictions on certain types of water use or efficiency standards for plumbing fixtures. It can also include changing management practices at the scale of individual enterprises. This includes changing such wasteful practices as irrigating during the day and using potable water for irrigation purposes, as well as shifting to more water-efficient crops and shifting industrial processes away from water-intensive production.

Demand Management by Sector

The different categories of policy instruments can be assessed in more detail if considered sectorally. Each sector—agricultural, industrial, and domestic—has a range of specific demand-management options open to it, ranging from technological changes to changes in management practices.

Agricultural Sector

Improving agricultural water productivity. Changing patterns of demand, increased populations, and poverty and food insecurity in marginalized communities, exacerbated by changes in rain patterns (e.g., higher variability, increased or decreased total rainfall) due to climate change, are likely to require further productivity gains from agriculture and from the water that feeds it. This will take place as a result of agricultural water management practices implemented in a way that will improve equitable access and the conservation of the water resource base (FAO, 2003b). Water productivity in agriculture, defined as the water consumption per kilogram of output, is estimated to have doubled since the 1960s (FAO, 2003b). This is in part due to improved water management and conservation in both rainfed and irrigated agriculture, and partly because of the intensification attributable to fertilizer application and the use of high-yield varieties (FAO, 2003b). By one calculation, a 10-percent increase in water productivity would save the same amount of water as current domestic water consumption, making investing in water management in agriculture an attractive option for water managers looking to free water for use in other sectors (FAO, 2003b).

Water productivity can be improved by doing four things: increasing the marketable yield of a crop for each unit of water transpired by it; changing the variety of crops grown; reducing drainage, seepage, percolation, and evaporative outflows; and increasing the effective use of alternative water resources such as rain, stored water, and water of marginal quality (FAO, 2003b). These actions can be taken at different scales, from plant to basin-wide levels. One example is the reuse of urban wastewater as irrigation water. Depending on its first use and its intended reuse, the water may be treated or used directly. The water might be discharged into wetlands for treatment, and then drawn for use as irrigation water.

The agricultural sector has water requirements not only for crops, but also for livestock and fish. Because of the relatively small volume of water required for these uses compared to crops, they are less discussed. However, with growing demand for animal products due to changes in eating habits with rising living standards, as well as increases due to population growth, this part of the agricultural sector may become more important. Aquaculture and livestock can produce a very high value of output for the water they use (Bakker, et al., 1999; Meinzen-Dick and Appasamy, 2002), and, much like potential shifts to industry from agriculture in the face of reduced water supplies, there may also be a shift to these kinds of agricultural products. There is, however, a risk of overgrazing, which can severely traumatize land and water resources, or, if grain-fed, the virtual water that is embedded in the food fed to the livestock. There may be a local reduction in water resources used, but perhaps not an overall reduction. This may be, however, a way to shift resources from areas with ample supplies to those without.

Irrigation efficiency improvements. There are several avenues by which the irrigation of crops may be modified to reduce the agricultural sector's demand for water: irrigation efficiency improvements, the reuse of urban wastewater in agriculture, and the reduction or elimination of subsidies for irrigation water.

Irrigation efficiency is a measure of the amount of water required to irrigate a field, farm, or watershed. An efficient irrigation system can irrigate a crop with a minimum of waste or losses. In agriculture, irrigation efficiency is low. In the western United States, efficiencies are at about 55 percent; in the Indus region of Pakistan, they are less than 40 percent; and in many other countries they are also less than 50 percent. Although there is much room for improvement, the figures can be somewhat misleading. The water that one farm does not use may be used by the next farm downstream, raising the apparent system efficiency (Gleick, 2000a). Nonetheless, there is much room for more efficient irrigation technologies.

Irrigation water is lost in several ways. As Figure 6.1 shows, 55 percent of irrigation water is not used effectively by crops. In general, about 45 percent of the water losses occur when the water is applied to fields, while the rest is lost in equal amounts to transmission to and during use on the farm. The most successful efficient technology has been drip irrigation, which replaces the flooding of fields with the precise application of water at the roots of plants, resulting in large reductions in evaporative losses as well as productivity gains, sometimes as large as 200 percent or more (Gleick, 2000b). Other technological improvements include laser leveling of fields and advanced drainage systems that reduce salinization. These technologies can raise irrigation efficiencies from 60 percent to 95 percent (Wolff and Gleick, 2002).

These technologies are relatively expensive, and poorer farmers do not generally have the resources to pay for them. In these cases, traditional techniques can also save water and improve productivity.

Figure 6.1
Water Losses in Irrigation

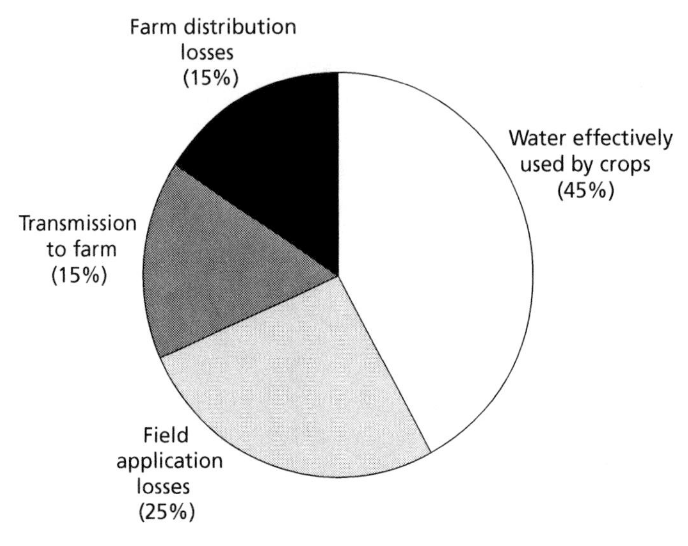

SOURCE: FAO (2003a).
RAND *MG358-6.1*

These technologies include water-harvesting techniques, using rocks to slow runoff on terraced land, and using small reservoirs to store rainwater from the rainy season to be used for crops in the dry season (Gleick, 2001b).

It is important, however, to investigate fully the implications of all such technologies. Increases in water productivity do not necessarily result in greater economic or social benefits (FAO, 2003b). Water has many uses in rural parts of developing countries, such as the production of timber firewood and fiber, raising fish and livestock, and domestic and environmental uses. The water is both a public and a social good, and the reduction of irrigation seepage or runoff may impair these other uses. For example, in Sri Lanka, a leaky irrigation project was found to be the source of sustenance for agroforestry that was important socially and economically for the local community (FAO, 2003b). In arid parts of Mali and Burkina Faso, similar gains are realized from agroforestry components of irrigation projects.

These examples show the importance of understanding the multiple roles of water in a community before proceeding with any projects.

Water prices. Releasing water from the agricultural sector frees it for other sectors, or for further irrigation projects, thereby reducing conflicts and avoiding development of new supplies (El-Ashry, 1995). Agricultural productivity would improve at the same time. However, the current management of water in most places provides far more disincentives than incentives for conservation in agriculture. Low water prices are one of the most visible disincentives to investment in conservation technology and management expertise (El-Ashry, 1995). Huge publicly funded price subsidies are common in most countries, including the United States and Mexico, where users pay about 11 percent of the full cost of water (El-Ashry, 1995). This encourages producers to grow crops that are both low value and water intensive, and removes any incentive to use water efficiently (Gleick, 2000a).

Salinization. Poor management of irrigation systems has resulted in inefficiencies, poorer than expected crop yields, and salinization problems that have affected productivity on as much as 20 percent of irrigated lands, or about three percent of total cultivated land worldwide (Ghassemi, Jakeman, and Nix, 1995). About 22 percent of irrigated land in the United States is affected, as is almost 39 percent in Argentina, 33 percent in Egypt, 30 percent in Iran, 26 percent in Pakistan, 17 percent in India, and 14 percent in China (Ghassemi, Jakeman, and Nix, 1995). This problem persists anywhere large-scale irrigation is practiced. There are disincentives for farmers to reduce the pollution caused by the application of chemicals to their crops. Drainage water from many agricultural areas contains toxic levels of chemicals for wildlife. Because irrigators do not feel the impact of this runoff, they are not motivated to fix the problem.

Food imports and storage. In lieu of the expansion of irrigated agriculture or importation of water, it may be more economically viable for water-short countries to import water-intensive food products from water-surplus areas. Food can be thought of as virtual water, and is a relatively efficient means of moving water, since the product is perhaps a thousandth of the weight of the water needed to grow it (Zaba and Madulu, 1998). Along with the water savings enjoyed by the country importing the food, there may be global water savings due to the greater water productivity in the producing versus the importing country. For example, corn grown in Europe uses approximately 0.6 m^3 water per kilogram of corn, versus 1.2 m^3 per kilogram in Egypt, a savings of 0.6 m^3 of water for every kilogram of corn imported by Egypt instead of grown there (FAO, 2003b). Additionally, the water that is used in the growing of the imported crop may be renewable in the exporting country, but would be nonrenewable in the importing country. For example, wheat grown in Saudi Arabia uses fossil groundwater, unlike that grown in parts of Europe (Smith, 2003).

Imports of food assume a surplus in other parts of the world and that the importing country has the income required for imports. In many countries where water and food are in short supply, the poorly developed economy provides low purchasing power. The situation is worsened by the fact that logistical problems, trade barriers, and political and social instabilities encumber international trade or food-relief programs. Also, an emphasis on national food security for political, cultural, or other reasons impairs the rational trade of food, and hence a more rational use of water. A reliable global food security system would allow water-scarce countries to target their water for domestic and industrial use while importing food (WWC, 2000).

Some plants and animals provide relatively little nutritional value per amount of water required. In fact, another method of conserving water is to choose different foods to eat, since, for example, a pound of beef takes 15 to 30 times more water to produce than a pound of corn, and three to 12 times more water than a pound of chicken.

Another water-saving tactic in arid and semi-arid countries is food storage. If surplus production in a wet year is stored, it can be used during dry years where yield is reduced. Because water productivity is higher in the surplus year, real water savings can be realized. This tactic has been practiced since biblical times. More recently, in Syria in the years 1988 and 1989, some 2.8 billion m^3 of water was saved by storage of food (FAO, 2003b).

Domestic Sector

Many municipalities have implemented successful conservation programs, achieving reductions in water demand ranging from 10 to 50 percent (Wolff and Gleick, 2003; Postel, 1997). They have achieved these successes using integrated water-resource management, including such tools as monitoring and evaluation of water use; indoor and outdoor efficiency standards; technical assistance programs; education programs such as demonstration gardens, landscape seminars, and school education programs; economic incentives such as conservation rate structures, rebates for decentralized investments; and full economic integration of efficiency improvements when developing new water supplies (Wolff and Gleick, 2003).

Reducing losses from nonrevenue water. A large proportion of the water wasted in the municipal and domestic sector is in the delivery of water to and within towns. Much like losses associated with irrigation systems, cities lose from three percent to 70 percent of their water as nonrevenue water (NRW) or unaccounted-for water (UFW) (Table 6.1). NRW is the difference between the quantity of water supplied to a city's network and the metered quantity of water used by customers. Rates of NRW are far higher in developing countries, as is intimated in Table 6.1. Throughout Latin America, losses of 40 to 70 percent are common (WHO, 1992). In India, 35 urban centers average 26 percent of water lost, while Asian cities have losses ranging from four to 65 percent (McIntosh, 2003). This can be compared with Geneva, Switzerland, where losses have been reduced to 13 percent, or Singapore, where losses are six percent (WHO, 1992; McIntosh, 2003). NRW tends to be higher in areas where piped water coverage is lower.

NRW has three components: (a) real losses (physical losses due to leakage and overflow from pipes and reservoirs), (b) apparent losses (administrative losses due to illegal connections and inaccuracies of water meters), and (c) unbilled authorized consumption (water for

Table 6.1
Nonrevenue Water, by Region

Region	Percent of Supply
Africa	39
Asia	30
Latin America and the Caribbean	42
North America	15

SOURCES: WHO/UNICEF (2000); McIntosh (2003).

firefighting, main flushing, and process water for waste-treatment plants, for example). While every case is different, often real losses and apparent losses contribute roughly equally to UFW, while unbilled authorized consumption is a minor component of losses (Saghir, Schiffler, and Woldu, 2000). There are three well-known and technically uncomplicated steps toward solving this problem:

- Reduce physical losses to the lowest economically feasible level.
- Meter at least all major consumers (universal metering may have to be a longer-term project, since meters are a major foreign-currency expenditure in developing countries).
- Bill everyone for water supplied, and enforce payment (Rogers, Kalbermatten, and Middleton, 1999).

The benefits of reducing NRW, especially in developing countries, include the following:

- allowing investments in new works to be deferred or at least reduced in scope, with significant savings
- greatly increased revenue to pay for water treatment and distribution as well as operation and maintenance when the meter-reading and billing system and the detection and billing of illegal connections are improved. For example, in urban areas in Thailand in the 1980s, each 10 percent of unaccounted-for water saved was estimated to generate immediately an additional

$8 million per year from the 3.5 million people served (WRI, 1997).

- reduction in wasteful consumption, which has been shown to be connected to metering and adequate water rates (McIntosh, 2003). This will likely lead to decreases in total consumption.
- improved understanding of consumption patterns, allowing the optimization of distribution systems and improved demand projections (McIntosh, 2003)
- reduced wastewater and pollution.

These benefits depend not only on improved management of water losses, but also on reasonable pricing of water and water services. If water customers pay low tariffs, there is little incentive for utilities to reduce apparent losses and for customers to deal with leaks and wasteful use.

Improved household and municipal water efficiency can also be attained, especially by using water-saving devices. In developed countries, toilet flushing is the largest indoor use of water in single-family homes, and one flush can use as much water as an entire day's worth of water in a rural village in a low-income country (Gleick, 2000b). Many municipalities have revised their building codes to require low-flow toilets, faucets, and showers. These devices can save large amounts of water at very low cost. Some developing countries are following suit. In Mexico City, for example, 350,000 six-gallon-per-flush toilets were replaced by 1.5-gallon-per-flush toilets, saving enough water to supply 250,000 additional residents (Gleick, 2001b).

Toilet flushing, via new technology, can even be eliminated. New toilets manage human waste without the use of water in a convenient, safe, and odor-free way. Although they are still expensive and unfamiliar to most consumers, their existence points to new ways of conserving water.

Pricing. In the domestic sector, it is argued that wasteful household practices such as watering thirsty lawns and maintaining swimming pools can be reduced through pricing. However, the demand for water is inelastic for some uses over a wide range of quantities (Brookshire and Whittington, 1993). In the domestic sector of developing countries, even poor households show willingness to pay for a reliable supply of water, as discussed earlier. However, it is thought that demand is only price inelastic in the short term, and that water users will change their behavior if the final, water-based service they desire is maintained (Wolff and Gleick, 2002).

Industrial Sector

In industry, the redesign of production processes to require less water per unit of production and the reuse and recycling of water in current production processes have resulted in large efficiency improvements in some industries. In steel production, new technology uses less than six cubic meters of water per ton of steel produced, as compared to the 60 to 100 m^3 used by the old technology. Aluminum can be made with 1.5 m^3 per ton produced, so even greater water savings can be realized by substituting aluminum for steel in the production process. Potential and actual savings in the United States have been estimated to range from 16 to 34 percent (Wolff and Gleick, 2002). Often the payback period for these conservation measures has been found to be short, as little as a year or less.

In the developing world there is even more scope for improvement. In China, steel making still uses 23 to 56 m^3 of water per ton of steel, and paper manufacturers use 900 m^3 of water per ton compared to the 450 m^3 per ton used in Europe. In all parts of the world, the valuing of water at market prices improves water-saving and efficiency measures undertaken by industry. Inexpensive water subsidizes inefficiency (de Villiers, 2000).

Demand management and supply management are both necessary for the efficient use of water resources. As seen here, they are multifaceted categories of management tools. As previously discussed, the combination of tools selected for a particular location will depend on its social, political, economic, and physical climate. Tools of different types are also applicable to different levels of management. In other words, a water system's manager and a small city's government will not have influence on food imports, but they can work toward improving local irrigation systems' efficiency or reallocating irrigation water to industry, for example. Other tools may be facilitated by policies at the state or national level, even if the choice of tools will be locally based. Whatever the case, these choices, and the mitigating factors illustrated in the framework at the beginning of this study, will determine the ultimate sustainability of water resources.

CHAPTER SEVEN

Conclusions: The Water Crisis Revisited

The introductory chapter of this study considered the idea of a worldwide water crisis brought on by increasing global population, and concluded that studies arguing that such a crisis is imminent are probably alarmist and overreaching the predictive ability of the available data. Instead, it was argued that a different model of interaction between demographic variables and water resources must be used to examine the future of water worldwide.

The preferred model includes demographic factors and water resource variables, with myriad factors dampening or heightening the relationship between these variables (see Figure 7.1). An examination of these factors together, rather than one or two alone, can give a far better indication of the potential for water crisis on a global scale. It is also important to include data that are local in nature, whenever possible. Often, however, these data are not available, and until they are, the quality of predictions will suffer.

Demographic factors work together to place stress on water resources, both by increasing demand in many places and by reducing supply through pollution and changes in environmental conditions. In this monograph, six demographic trends—population growth, number of households, urbanization, migration, population distribution, and mortality—have been identified that affect demand for water resources, two of which—mortality and migration—are themselves affected by water quality and quantity (Figure 7.1). In addition,

Figure 7.1
Model for Analyzing Interactions Between Demographic Factors and Water Resources

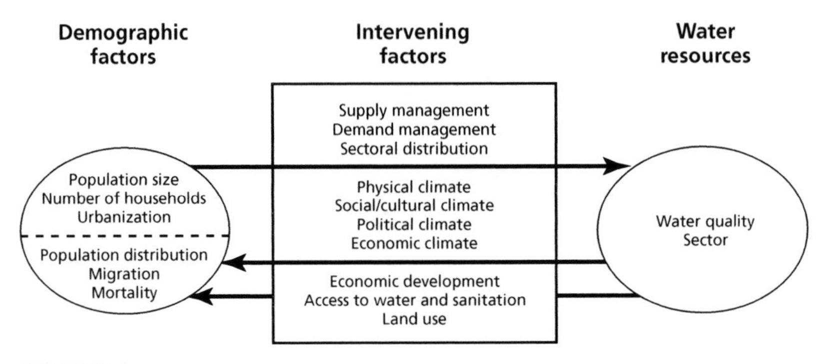

RAND *MG358-7.1*

several factors can be used to understand and manage the water resource. These are necessary to grasp better how supply of and demand for water is structured and what influences supply and demand (beyond demographic variables), and subsequently to adjust supply and demand via management and policy tools to achieve sustainability.

Demographic factors affect freshwater systems in various direct and indirect ways. Population growth, changes in household size, and urbanization influence land-use changes such as deforestation and land clearance for building and agriculture, which encourage pollution and siltation of water supply systems. Pollution of water systems is increased when chemicals, nutrients, and other wastes are released into the atmosphere from industrial, domestic, and agricultural practices, which are also impacted by population growth, urbanization, and societal and individual income levels. Population growth and urbanization can also lead to the overextraction of water from water systems. Pollution, land clearance, and overextraction all can permanently damage and destroy water systems, reciprocally affecting human systems.

Many other *social, cultural, and physical pressures* are exacted on the systems, and the combination of all of these pressures can potentially create conditions of scarcity or crisis. The damage and destruction of water systems reduces the supply of water, exacerbating any existing water supply problems, while increasing mortality and morbidity, and influencing migration decisions in populations that are exposed to these changes. Although water scarcity may be rooted in genuine shortages, it may also be a social construct, crafted from expectations, and exacerbated by lack of knowledge or human resources. As discussed, scarcity has several causes of both natural and human agency, all of which interact and influence one another, and several of which are due to the direct or indirect impact of factors such as changes in demographic variables, changes in land use, polluting of water supplies, financial and institutional failures, and climatic conditions and variability.

The effective management of water systems requires the knowledge and understanding of intervening factors to the sustainability of water systems. For example, this study has pointed out the importance of the *sectoral distribution of water* in the management of water systems. Although it often gets the most attention, both politically and in reports, the domestic and municipal sector, in most places, uses the least amount of water. This is partly because of its importance to human health, and because of the local impacts of urbanization. Nonetheless, an overly exclusive focus on the domestic sector can derail attempts to balance overall supply and demand for a country or watershed.

Sectoral demand for water is also important to how demographic and economic factors will affect future demand for water. The demand for water in the domestic sector, for example, is most influenced by the number of households, the population size, and urbanization levels, as well as by the level of economic development. Demand for water in this sector tends to increase with rising incomes, rising numbers of households, and rising urbanization rates. Levels of economic development and of urbanization affect demand in the industrial sector. The level of economic development affects agricultural-sector demand; per-capita demand tends to decrease with rising income levels. However, climate and culture can overcome all of these effects.

Knowledge of the distribution of and predicted changes in these demographic factors will help to inform managers and policymakers about changes in demand in different sectors, and therefore help in making choices in demand and supply management policies. Since each sector demands water in both urban and rural areas, all must be considered when making these choices, including the poor (who make up about one-third of the global population), who may be suffering the greatest consequences of poor access to water and sanitation.

Every watershed—and locales within watersheds—has its own set of physical and social profiles that affect water supply and demand within that area. Local areas have differences that are sectoral, geographical, environmental, cultural, economic, and institutional. Demographic variables affect many of these local differences, just as they do at the larger scale. Ideally, the amount of water available to each locale would be known. But even without these data, much information can be gathered to inform managers. This information can be gathered and changes mostly easily accomplished, with the cooperation of those who will see the effects of the decisions. This underlines the importance of including communities and individuals when making decisions concerning water availability and use.

In providing for water for communities and for the maintenance of natural systems into the future, managers have tools to both reduce demand and augment supply. The combination of tools chosen will depend on the place and on the conditions there. Demand management is appealing, especially in countries with few financial resources, because it can help to defer large investments in infrastructure, essentially buying time. A $2 million investment in demand management could delay a new investment of $100 million for 10 years (UNCHS, 2000). However, lack of storage in these systems can be a severe constraint, and needs for supply must also be met, including urban and rural water and sanitation infrastructure, which, among other benefits, will help to maintain supply by reducing the pollution of water sources.

Ultimately, it is the interplay between the natural systems supplying water and the human systems demanding water that is important to the sustainability of water resources. For that reason, and because social and economic stability are linked to the provision of water to urban, peri-urban, and rural consumers in all sectors, *the governance of water systems is an extremely important part of sustaining water systems into the future.* This governance may best take the form of a broad-based partnership of public, private, and community sectors. The private sector may contribute efficiency gains in water management as well as financing. Communities are important for their contributions to transparency, equity, sense of ownership, local knowledge of cultural and social conditions, and to help in cost recovery. The public sector is most important for its role in policy setting and as a regulatory agency, and as a source of funding for new infrastructure in some areas. Such a partnership may have the best chance to provide water efficiently, equitably, and in a manner that can be sustained into the future.

Will there be a global water crisis? There will undoubtedly continue to be localized problems, and perhaps more widespread problems in some areas, depending on local physical, social, economic, and cultural conditions. However, a global water crisis can be averted. A society confronting scarcity generally has choices. An exception may be those in a state of water poverty defined by a relative lack of both natural and social resources. There are many options for improving water management and alternatives for meeting supply and demand even for growing and changing populations. Some of these solutions are expensive, and others face serious infrastructural obstacles in some countries. For this reason and many others, attention to demographic factors is an important part of the formula for staving off water crisis indefinitely. To be successful, there must be sufficient institutional, intellectual, and administrative capacity to create solutions and to carry through actions. For countries without these resources, additional help will be needed to ensure that all people, no matter how many, realize their basic right to clean, adequate supplies of water. To achieve that, sustainable water development and management requires the integration of social and economic concerns with environmental concerns. This effort will be enhanced by research that focuses on a scale as small as possible, to provide data and information on methods that will help to manage water at socially relevant levels, more locally, as well as at environmentally relevant levels such as the watershed. In this vein as well, research focusing on demographic variables that are less understood, such as the impact of the number and size of households on resources and the environment, will help to improve the overall understanding of the relationship between demography and water resources.

References

Abramovitz, Janet N., and Jane A. Peterson, *Imperiled Waters, Impoverished Future: The Decline of Freshwater Ecosystem*, Worldwatch Paper 128, Washington, D.C.: World Resource Institute, 1996.

Aggarwal, Rimjhim, Sinala Netanyahu, and Claudia Romano, "Access to Natural Resources and the Fertility Decision of Women: The Case of South Africa," *Environment and Development Economics*, Vol. 6, No. 2, 2001, pp. 209–236.

Altaf, Mir Anjum, Dale Whittington, Haroon Jamal, and V. Kerry Smith, "Rethinking Rural Water Supply Policy in the Punjab, Pakistan," *Water Resources Research*, Vol. 29, No. 7, 1993, pp. 1943–1954.

Association for International Water and Forest Studies, *Power Conflicts: Norwegian Hydropower Developers in the Third World*, 1994.

———, *Don't Dam(m)it! Impacts of Large Hydropower Projects in the Third World*. Online at http://www.fivas.org/english/dumbdams.htm as of April 8, 2005.

Austin, L. M., and S. J. van Vuuren, *Sanitation, Public Health and the Environment: Looking Beyond Current Technologies*, 2002. Online at http://www.csir.co.za/websource/ptl0002/docs/boutek/akani/2002/mar/auspaper1.pdf as of April 7, 2005.

Baer, A., "Not Enough Water to Go Around?" *International Social Science Journal*, Vol. 48, No. 148, 1996, pp. 277–292.

Bakker, Margaretha, Randolph Barker, Ruth Meinzen-Dick, and Flemming Konradsen, eds., *Multiple Uses of Water in Irrigated Areas: A Case Study from Sri Lanka*, SWIM Report 8, Colombo, Sri Lanka: International

Water Management Institute, 1999. Online at http://www.cgiar.org/iwmi/pubs/SWIM/Swim08.pdf.

Barlow, Maude, and Tony Clarke, "Who Owns Water?" *The Nation*, September 2, 2002. Online at http://www.thenation.com/doc.mhtml?i=20020902&s=barlow as of April 8, 2005.

Bates, Diane C., "Environmental Refugees? Classifying Human Migrations Caused by Environmental Change," *Population and Environment*, Vol. 23, No. 5, 2002, pp. 465–477.

Bernstein, Stan, "Freshwater and Human Population: A Global Perspective," in Karin Krchnak, ed., *Human Population and Freshwater Resources: U.S. Cases and International Perspectives*, Yale School of Forestry and Environmental Studies Bulletin Series Number 107: New Haven, Connecticut: Yale University, 2002, pp. 149–157. Online at http://www.yale.edu/environment/publications/ as of April 5, 2005.

Bhatia, Ramesh, Rita Cestti, and J. T. Winpenny, *Water Conservation and Reallocation: Best Practice Cases in Improving Economic Efficiency and Environmental Quality*, Washington, D.C.: World Bank, 1995.

Biddlecom, Ann E., William G. Axinn, and Jennifer S. Barber, "Environmental Effects on Family Size Preferences and Subsequent Reproductive Behavior in Nepal," *Population and Environment*, Vol. 26, No. 3, 2005, pp. 583–621.

Briscoe, J., *Water Resources Management in Yemen: Results of a Consultation*, office memorandum, Washington. D.C.: World Bank, 1999, as quoted in Shah, 2000.

Brookshire, David S. and Dale Whittington, "Water Resources Issues in the Developing Countries," *Water Resources Research*, Vol. 29, No. 7, 1993, pp. 1883–1888.

Brown, Ellen P. and Robert Nooter, *Successful Small-Scale Irrigation in the Sahel*, World Bank Technical Paper No. 171, Washington, D.C.: World Bank, 1992.

Buckley, Richard, *The Battle for Water: Earth's Most Precious Resource*, Cheltenham, England: European Schoolbooks, 1994.

Butz, W. P., J. P. Habicht, and J. DaVanzo, "Environmental Factors in the Relationship Between Breastfeeding and Infant Mortality: The Role of

Sanitation and Water in Malaysia," *American Journal of Epidemiology*, Vol. 119, No. 4, 1984, pp. 516–525.

CEQ. See Council on Environmental Quality.

Cernea, Michael M.,"Involuntary Resettlement and Development," *Finance and Development*, Vol. 25, No. 3, 1988, pp. 44–46.

Cheng, Tony, "Million Moved by Chinese Dam," *BBC News: World: Asia-Pacific*, December 17, 2002. Online at http://news.bbc.co.uk/1/hi/world/asia-pacific/2579663.stm as of April 8, 2005.

Commission on Sustainable Development, United Nations Economic and Social Council, "Water: A Key Resource for Sustainable Development," 2001.

Cosgrove, William J. and Frank R. Rijsberman, *World Water Vision: Making Water Everybody's Business*, London: Earthscan Publications, 2000.

Council on Environmental Quality, *Environmental Quality: The World Wide Web: The 1997 Annual Report of the Council on Environmental Quality*, 1997. Online at http://ceq.eh.doe.gov/nepa/reports/1997/index.html as of April 11, 2005.

CSD. See Commission on Sustainable Development.

Dasgupta, Partha, "Population and Resources: An Exploration of Reproductive and Environmental Externalities," *Population and Development Review*, Vol. 26, No. 4, 2000, pp. 643–689.

de Villiers, Marq, *Water: The Fate of Our Most Precious Resource*, New York: Houghton Mifflin, 2000.

Dixon, Lloyd S., Nancy Y. Moore, and Susan W. Schechter, *California's 1991 Drought Water Bank: Economic Impacts on the Selling Regions*, Santa Monica, Calif.: RAND Corporation, MR-301-CDWR/RC, 1993.

"A Dry Future? India Will Be Water Stressed by 2025," *IEI News*, Vol. 67, 2000.

Dudley, Nigel, and Sue Stolton, *Running Pure: The Importance of Forest Protected Areas to Drinking Water: A Research Report*, Gland, Switzerland: WWF, 2003. Online at http://lnweb18.worldbank.org/ESSD/envext.nsf/80ByDocName/RunningPureTheimportanceofforestprotectedareastodrinkingwater/$FILE/RunningPure2003+.pdf as of April 7, 2005.

EEA. See European Environment Agency.

Economist, "Water," October 30, 1999.

ECSP. See Environmental Change and Security Project.

Edmonds, Sarah, "Canada Seeks to Shut Floodgates on Water Exports," *Planet Ark*, December 15, 1998. Online at http://www.planetark.org/dailynewsstory.cfm?newsid=1990&newsdate=15-dec-1998 as of April 8, 2005.

EEA. See European Environment Agency.

El-Ashry, Mohamed T., "Issues in Managing Water Resources in Semiarid Regions," *GeoJournal*, Vol. 35, No. 1, 1995, pp. 53–57.

Environmental Change and Security Project, *Finding the Source: The Linkages Between Population and Water*, New York: ECSP, 2002. Online at http://wwics.si.edu/index.cfm?topic_id=1413&fuseaction=topics.publications&group_id=6974 as of May 5, 2005.

European Environment Agency, *Europe's Environment: The Dobrís Assessment*, Copenhagen, Denmark: European Environment Agency, 1995. Online at http://reports.eea.eu.int/92-826-5409-5/en as of May 24, 2005.

———, "Households: Household Number and Size," *European Environment Agency: Information for Improving Europe's Environment*, Copenhagen, Denmark: European Environment Agency, 2001. Online at http://themes.eea.eu.int/Sectors_and_activities/households/indicators/consumption?printable=yes as of May 24, 2005.

Evison, G., *Child Mortality and Water Use in Mwanza Region, Tanzania*, unpublished MSc Dissertation, CPS, London School of Hygiene and Tropical Medicine, 1996.

Falkenmark, Malin, "Population, Environment and Development," in *United Nations, Population, Environment and Development: Proceedings of the United Nations Expert Group Meeting on Population, Environment and Development, United Nations Headquarters, 20–24 January 1992*, New York: United Nations, 1994, pp. 99–116.

Falkenmark, Malin, and Carl Widstrand, "Population and Water Resources: A Delicate Balance," *Population Bulletin*, Vol. 47, No. 3, 1992, pp. 1–36.

FAO. See Food and Agriculture Organization of the United Nations.

Feachem, R., "The Water and Sanitation Decade," *Journal of Tropical Medicine and Hygiene*, Vol. 84, No. 2, 1981.

Filmer, Deon, and Lant H. Pritchett, "Environmental Degradation and the Demand for Children: Searching for the Vicious Circle in Pakistan," *Environment and Development Economics*, Vol. 7, No, 1, 2002, pp. 123–146.

FIVAS. See Association for International Water and Forest Studies.

Food and Agriculture Organization of the United Nations, "The Contribution of Blue Water and Green Water to the Multifunctional Character of Agriculture and Land. Background Paper 6: Water," in *Cultivating Our Futures*, Maastricht, The Netherlands: Food and Agriculture Organization, 1999. Online at http://www.fao.org/docrep/x2775e/x2775e08.htm as of April 6, 2005.

———, "General Summary Asia: Water Withdrawal," *AQUASTAT: FAO's Information System on Water and Agriculture*, 2003a. Online at http://www.fao.org/ag/agl/aglw/aquastat/regions/asia/index4.stm as of April 6, 2005.

———, *Unlocking the Water Potential of Agriculture*, Rome: Food and Agriculture Organization, 2003b.

Fort, Denise D., "Water and Population in the American West," in Karin Krchnak, ed., *Human Population and Freshwater Resources: U.S. Cases and International Perspectives*, Yale School of Forestry and Environmental Studies Bulletin Series Number 107, New Haven, Conn.: Yale University, 2002, pp. 17–24. Online at http://www.yale.edu/environment/publications/bulletin/107pdfs/107Fort.pdf as of April 11, 2005.

Foster, Stephen S.D., Adrian Lawrence, and Brian Morris, *Groundwater in Urban Development: Assessing Management Needs and Formulating Policy Strategies*, World Bank Technical Paper, No. 390, Washington, D.C.: The World Bank, 1998.

Frederick, Kenneth D., "Marketing Water: The Obstacles and the Impetus," *Resources*, Vol. 132, 1998, pp. 7–10. Online at http://www.rff.org/rff/Documents/RFF-Resources-132-water.pdf as of August 12, 2005.

Frederick, Kenneth D., and Peter H. Gleick, *Water and Global Climate Change: Potential Impacts on U.S. Water Resources*, Arlington, Va.: Pew Center on Global Climate Change, 1999.

Frederick, Kenneth D., Jean Hanson, and Tim VandenBerg, *Economic Values of Freshwater in the United States*, Washington, D.C.: Resources for the Future, 1996.

Gardner, Robert and Richard Blackburn, "People Who Move: New Reproductive Health Focus," *Population Reports, Series J*, No. 45, 1996. Online at http://www.infoforhealth.org/pr/j45edsum.shtml as of April 7, 2005.

Gardner-Outlaw, Tom, and Robert Engelman, *Sustaining Water, Easing Scarcity: A Second Update*, Washington, D.C.: Population Action International, 1997.

Ghassemi, F., A. J. Jakeman, and H. A. Nix, *Salinisation of Land and Water Resources: Human Causes, Extent, Management and Case Studies*, Wallingford, Oxon, UK: CAB International, 1995, in Gleick, 2000b.

Giles, Harry, and Bryan Brown, "And Not a Drop to Drink: Water and Sanitation Services to the Urban Poor in the Developing World," *Geography*, Vol. 82, No. 2, 1997, pp. 97–109.

Gleick, Peter H., "An Introduction to Global Fresh Water Issues," in Peter H. Gleick, ed., *Water in Crisis: A Guide to the World's Fresh Water Resources*, New York: Oxford University Press, 1993, pp. 3–12.

———, "Basic Water Requirements for Human Activities: Meeting Basic Needs," *Water International*, Vol. 21, No. 2, 1996, pp. 83–92.

———, "The Changing Water Paradigm: A Look at Twenty-First Century Water Resources Development," *Water International*, Vol. 25, No. 1, 2000a, pp. 127–138. Online at http://www.iwra.siu.edu/win/win2000/win03-00/gleick.pdf as of April 5, 2005.

———, *World's Water 2000–2001: The Biennial Report on Freshwater Resources*, Washington, D.C.: Island Press, 2000b.

———, "Global Water: Threats and Challenges Facing the United States: Issues for the New U.S. Administration," *Environment*, Vol. 43, No. 2 2001a.

———, "Making Every Drop Count," *Scientific American*, Vol. 284, No. 2, 2001b, pp. 40–45.

———, *The World's Water 2002–2003: The Biennial Report on Freshwater Resources*, Washington, D.C.: Island Press, 2002.

Gumbo, B., "Non-Waterborne Sanitation and Water Conservation," in *Encyclopedia of Life Support Systems*, Oxford: EOLSS Publishers, 2001.

Hinrichsen, Don, "A Human Thirst," *World Watch*, Vol. 16, No. 1, 2003, pp. 12–18.

Hinrichsen, Don, Karin Krchnak, and Katie Mogelgaard, *Population, Water and Wildlife: Finding a Balance*, Washington, D.C.: National Wildlife Federation, 2002.

Hinrichsen, Don, Bryant Robey, and Ushma D. Upadhyay, "Solutions for a Water-Short World," *Population Reports*, Vol. M, No. 14, 1997. Online at http://www.infoforhealth.org/pr/m14edsum.shtml as of April 7, 2005.

Hunter, Lori M., *The Environmental Implications of Population Dynamics*, Santa Monica, Calif.: RAND Corporation, MR-1191-WFHF/RF/DLPF, 2000.

IJHD. See International Journal of Hydropower and Dams.

Institute of Water Research, *Introductory Land and Water Learning Module*, Michigan State University, 1997. Online at http://www.iwr.msu.edu/edmodule/welcome.htm as of April 6, 2005.

International Journal of Hydropower and Dams, *1998 World Atlas and Industry Guide*, Surrey, UK: Aqua-Media International, 1998.

International Water Management Institute, "Overview: Sustainable Groundwater Management Theme," *Research Themes: Groundwater*, 2001. Online at http://www.cgiar.org/iwmi/groundwater/index.htm as of April 6, 2005.

IUCN. See World Conservation Union.

IWMI. See International Water Management Institute.

Krock, Lexi, "Miracle of Rice," *NOVA Online: Japan's Secret Garden*, 2000. Online at http://www.pbs.org/wgbh/nova/satoyama/rice.html as of April 8, 2005.

Lenzen, Manfred, "The Influence of Lifestyles on Environmental Pressure," *Year Book Australia 2002: Environment*, 2002. Online at http://www.abs.gov.au/Ausstats/abs@.nsf/0/372480889969a1d9ca256b35007ace09?OpenDocument.

Liu, Jianguo, Gretchen C. Daily, Paul R. Ehrlich, and Gary W. Luck, "Effects of Household Dynamics on Resource Consumption and Biodiversity," *Nature*, Vol. 421, No. 6922, 2003, pp. 530–533.

Lobina, Emanuele, and David Hall, *Problems with Private Water Concessions: A Review of Experience*, London: Public Services International Research Unit (PSIRU), 2003. Online at http://www.psiru.org/reports/2003-06-w-over.doc as of April 8, 2005.

Lutz, W., W. Sanderson, and S. Scherbov, "The End of World Population Growth," *Nature*, Vol. 12, 2001, pp. 543–546.

L'vovich, M. I., and G. F. White, "Use and Transformation of Terrestrial Water Systems," in Billie Lee Turner, William C. Clark, Robert W. Kates, John F. Richards, Jessica T. Mathews, and William B. Meyer, ed., *The Earth as Transformed by Human Action: Global and Regional Changes in the Biosphere over the Past 300 Years*, Cambridge, UK: Cambridge University Press, 1991, pp. 235–252.

MacKellar, F. Landis, Wolfgang Lutz, Christopher Prinz, and Anne Goujon, "Population, Households, and CO2 Emissions," *Population and Development Review*, Vol. 21, No. 4, 1995, pp. 849–865.

Maita, H., "Water Resources, Forests, and Their Related Issues in Japan," *Proceedings of 2003 Tsukuba Asian Seminar on Agricultural Education*, University of Tsukuba, Japan, 2003.

Martin, Nicola, *Population, Households and Domestic Water Use in Countries of the Mediterranean Middle East (Jordan, Lebanon, Syria, the West Bank, Gaza and Israel)*, International Institute for Applied Systems Analysis (IIASA), Interim Report IR-99-032,1999. Online at http://www.iiasa.ac.at/Publications/Documents/IR-99-032.pdf as of April 7, 2005.

Martindale, Diane, "Sweating the Small Stuff: Extracting Freshwater from the Salty Oceans Is an Ancient Technique That Is Gaining Momentum in a High-Tech Way," *Scientific American*, 2001, pp. 52–53. Online at http://www.internal.eawag.ch/~wehrli/exams/1_Safeguarding_water.pdf as of August 12, 2005.

McIntosh, Arthur C., *Asian Water Supplies: Reaching the Urban Poor*, London, UK: Asian Development Bank, 2003. Online at http://www.adb.org/documents/books/asian_water_supplies/default.asp as of April 8, 2005.

McKenzie, David. and Isha Ray, "Household Water Delivery Options in Urban and Rural India," Stanford Center for International Development, Working Paper No. 224, 2004. Online at http://scid.stanford.edu/pdf/SCID224.pdf as of April 8, 2005.

Meinzen-Dick, Ruth, and Paul P. Appasamy, "Urbanization and Intersectoral Competition for Water," in Woodrow Wilson International Center for Scholars Environmental Change and Security Project (ECSP), *Finding the Source: The Linkages Between Population and Water*, New York: ECSP, 2002. Online at http://wwics.si.edu/topics/pubs/popwawa4.pdf as of April 5, 2005.

Merrick, Thomas W., "The Effect of Piped Water on Early Childhood Mortality in Urban Brazil, 1970 to 1976," *Demography*, Vol. 22, No. 1, 1985, pp. 1–24.

Moench, Marcus, "Water and the Potential for Social Instability: Livelihoods, Migration and the Building of Society," *Natural Resources Forum*, Vol. 26, No. 3, 2002, pp. 195–204.

Murray, Christopher J. L., and Alan D. Lopez, *The Global Burden of Disease: A Comprehensive Assessment of Mortality and Disability from Diseases, Injuries, and Risk Factors in 1990 and Projected to 2020*, Cambridge, Mass.: Harvard School of Public Health on behalf of the World Health Organization and the World Bank, distributed by Harvard University Press, 1996.

Myers, Norman, "Environmental Refugees in a Globally Warmed World," *BioScience*, Vol. 43, No. 11, 1993, pp. 752–761.

National Council for Science and the Environment, *Report of the International Conference on Population and Development Cairo*, 1994. Online at http://www.cnie.org/pop/icpd/index.htm as of April 7, 2005.

Natural Resources Canada, "Weathering the Changes: Climate Change in Ontario," *Climate Change in Canada: Our Water*. Online at http://adaptation.nrcan.gc.ca/posters/articles/on_05_en.asp?Region=on&Language=en as of May 24, 2005.

NRC. See Natural Resources Canada.

Orians, Carlyn E. and Marina Skumanich, *The Population-Environment Connection: What Does It Mean for Environmental Policy?* Seattle: Battelle Seattle Research Center, 1995.

PAI. See Population Action International.

Patel, M., "Modelling Determinants of Health," *International Journal of Epidemiology*, Vol. 10, 1981, pp. 177–180.

Pebley, Anne R., "Demography and the Environment," *Demography*, Vol. 35, No. 4, 1998, pp. 377–389.

Platt, Rutherford H., "Protecting the New York City Water Supply Through Negotiated Watershed Agreement," *Remarks Presented to Institute for Civil Infrastructure Systems (ICIS) Workshop on Best Practices in Public Participation for Infrastructure Decision-Making*, Washington D.C. October 27, 2001. Online at http://www.icisnyu.org/admin/files/Rutherford%20Platt.pdf as of April 7, 2005.

Poff, N. LeRoy, Mark M. Brinson, and John W. Day, Jr., *Aquatic Ecosystems and Global Climate Change: Potential Impacts on Inland Freshwater and Coastal Wetland Ecosystems in the United States*, Arlington, Va.: Pew Center on Global Climate Change, 2002. Online at http://www.pewclimate.org/docUploads/aquatic.pdf as of August 12, 2005.

Population Action International, *Sustaining Water: Population and the Future of Renewable Water Supplies*, Washington, D.C.: Population Action International, 1993. Online at http://www.cnie.org/pop/pai/h2o-toc.html as of May 24, 2005.

Postel, Sandra, *Last Oasis: Facing Water Scarcity*, New York: W.W. Norton, 1997.

Pretty, Jules N., *Regenerating Agriculture: Policies and Practice for Sustainability and Self-Reliance*, Washington, D.C.: Joseph Henry Press, 1995. Online at http://www.nap.edu/books/0309052467/html/index.html as of April 8, 2005.

Puffer, Ruth Rice, and Carlos V. Serrano, *Patterns of Mortality in Childhood: Report of the Inter-American Investigation of Mortality in Childhood*, Washington, D.C.: Pan American Health Association, 1973.

Regmi, Anita, and John Dyck, "Effects of Urbanization on Global Food Demand," in Anita Regmi, ed., *Changing Structure of Global Food Consumption and Trade*, Market Trade and Economics Division, Economic Research Service, United States Department of Agriculture, Agriculture and Trade Report, WRS-01-1, 2001, pp. 23–30. Online at http://www.nal.usda.gov/fsrio/ppd/ers02.pdf.

Revenga, Carmen, Jake Brunner, Norbert Henninger, Richard Payne, and Ken Kassem, *Pilot Analysis of Global Ecosystems: Freshwater Systems*, Washington, D.C.: World Resources Institute, 2000. Online at http://pubs.wri.org/pubs_content.cfm?PubID=3056 as of April 5, 2005.

Revenga, Carmen, Janet Nackoney, Eriko Hoshino, Kura Yumiko, and Jon Maidens, *Watersheds of the World: Water Resources eAtlas*, Washington, DC: World Resources Institute, 2003. Online at http://www.iucn.org/themes/wani/eatlas/index.html as of April 11, 2005.

Ricciardi, Anthony, and Joseph B. Rasmussen, "Extinction Rates of North American Freshwater Fauna," *Conservation Biology*, Vol. 13, No. 5, 1999, pp. 1220–1222.

Rivera, Alfonso, Diana M. Allen, and Harm Maathuis, "Climate Variability and Change: Groundwater Resources," *Threats to Water Availability in Canada*, Burlington, Ont.: National Water Resources Institute, 2003. Online at http://www.nwri.ca/threats2full/ch10-1-e.html as of August 12, 2005.

Rock, M.T., "The Dewatering of Economic Growth," *Journal of Industrial Ecology*, Vol. 4, No. 1, 2000, pp. 57–73.

Rogers, Peter, *Water in the 21st Century: The Looming Crisis Averted?* paper presented at the 34th Annual Meeting of the Board of Governors of the Asian Development Bank, Honolulu, May 7, 2001. Online at http://www.adb.org/AnnualMeeting/2001/Seminars/rogers_paper.pdf as of April 8, 2005.

Rogers, Peter, John Kalbermatten, and Richard Middleton, *Water for Big Cities: Big Problems, Easy Solutions?* Comparative Urban Studies Project Policy Brief, 1999. Online at http://wwics.si.edu/index.cfm?topic_id=1410&fuseaction=topics.publications&doc_id=24946&group_id=24237 as of April 8, 2005.

Rosegrant, Mark W., "Water Resources in the Twenty-First Century: Challenges and Implications for Action," *A 2020 Vision for Food, Agriculture, and the Environment*, 2020 Discussion Paper No. 20, Washington, D.C.: International Food Policy Research Institute, 1997. Online at http://www.ifpri.org/2020/dp/dp20.htm as of April 8, 2005.

Saghir, Jamal, Manuel Schiffler, and Mathewos Woldu, *Urban Water and Sanitation in the Middle East and North Africa: The Way Forward*, Washington, D.C.: World Bank, 2000. Online at http://lnweb18.worldbank.org/mna/mena.nsf/f34b224d37365b3f852567ee0068bd93/2421f467c2c02626852569510066660e9/$FILE/way-english.pdf as of August 12, 2005.

Savenije, Hubert H.G., "The Role of Green Water in Food Production in Sub-Saharan Africa," FAO, 1999. Online at http://www.fao.org/ag/agl/aglw/webpub/greenwat.htm as of April 6, 2005.

Schneider, Keith, "Hitting the Bottle: Michigan Residents Fight for Control of the State's Water," *Grist Magazine*, October 23, 2002. Online at http://www.gristmagazine.com/maindish/schneider102302.asp as of April 8, 2005.

Schultz, T. P., "Interpretations of Relations Among Mortality, Economics of Household, and the Health Environment," in *Proceedings of the Meeting on Socioeconomic Determinants and Consequences of Mortality*, Mexico City, June 19–25, 1979, Geneva: World Health Organization, 1979.

Scientific American, *Science Desk Reference*, New York: John Wiley, 1999.

Serageldin, Ismail, *Toward Sustainable Management of Water Resources*, Washington, D.C.: World Bank, 1995.

Sexton, Richard, "The Middle East Water Crisis: The Making of a New Middle East Regional Order," *Capitalism, Nature, Socialism*, Vol. 3, No. 4, 1992, pp. 65–77.

Shah, Tushaar, *The Global Groundwater Situation: Overview of Opportunities and Challenges*, Columbo, Sri Lanka: International Water Management Institute, 2000.

Shaw, Roderick, "Re-use of Wastewater," in Roderick Shaw, ed., *Running Water: More Technical Briefs on Health, Water and Sanitation*, London, UK: Intermediate Technology, 1999.

Sheehan, Molly O'Meara, *Reinventing Cities for People and the Planet*, Washington, D.C.: Worldwatch Institute, 1999.

Shi, Anqing, *How Access to Urban Potable Water and Sewerage Connections Affects Child Mortality*, Washington, D.C.: World Bank, Policy Research Working Paper 2274, 2000.

Shiklomanov, Igor A., "Global Water Resources," *Nature and Resources*, Vol. 26, No. 3, 1990, pp. 34–43.

———, "World Fresh Water Resources," in Peter H. Gleick, ed., *Water in Crisis: A Guide to the World's Fresh Water Resources*, New York: Oxford University Press, 1993, pp. 13–24.

———, *Archive of World Water Resources and World Water Use*, Global Water Data files, State Hydrological Institute, St. Petersburg, Russia: CD-ROM, 1998.

———, "Appraisal and Assessment of World Water Resources," *Water International*, Vol. 25, No. 1, 2000, pp. 11–32. Online at http://www.iwra.siu.edu/win/win2000/win03-00/shiklomanov.pdf as of April 5, 2005.

Shogren, Elizabeth, "Sprawl Adds to Drought, Study Says," *Los Angeles Times*, August 29, 2002, p. A12.

Simmons, I.G., *Earth, Air, and Water: Resources and Environment in the Late 20th Century*, London: Edward Arnold, 1991.

SIWI. See Stockholm International Water Institute.

Smith, Craig S., "Saudis Worry As They Waste Their Scarce Water," *The New York Times*, January 26, 2003, p. 4.

Spaeth, Anthony, "Population Growth, Development, Bureaucracy: Bad Problems for Mother Earth," *TIME International*, Vol. 147, No. 13, 1996. Online at http://www.time.com/time/international/1996/960325/cover.indiaenviro.html as of May 24, 2005.

Squillace, Paul J., Michael J. Moran, Wayne W. Lapham, Curtis V. Price, Rick M. Clawges, and John S. Zogorski, "Volatile Organic Compounds in Untreated Ambient Groundwater of the United States, 1985–1995," *Environmental Science and Technology*, Vol. 33, No. 23, 1999, pp. 4176–4187.

Squillace, Paul J., Jonathon C. Scott, Michael J. Moran, B. T. Nolan, and Dana W. Kolpin, "VOCs, Pesticides, Nitrate, and Their Mixtures in Groundwater Used for Drinking Water in the United States," *Environmental Science and Technology*, Vol. 36, No. 9, 2002, pp. 1923–1930. Online at http://pubs.acs.org/cgi-bin/article.cgi/esthag/2002/36/i09/pdf/es015591n.pdf as of May 24, 2005.

Stockholm International Water Institute, *Water Management in Developing Countries: Policy and Priorities for EU Development Cooperation*, Stockholm: Stockholm International Water Institute, 2001. Online at http://www.siwi.org/downloads/Reports/Report%2012%20EU%20Communication%20Report%202001.pdf as of August 12, 2005.

Sutherland, Ben, "Dams Stir Water Arguments," BBC News World Edition, March 17, 2003. Online at http://news.bbc.co.uk/2/hi/science/nature/2856989.stm.

Sutherland, Elizabeth G., David L. Carr, and Siân L. Curtis, "Fertility and the Environment in a Natural Resource Dependent Economy: Evidence from Petén, Guatemala," presented at the Tercera Conferencia Internacional de la Poblacion del Istmo Centroamericano, November 17–19, 2003. Online at http://ccp.ucr.ac.cr/noticias/conferencia/pdf/sutherla.pdf as of April 7, 2005.

Swain, Ashok, "Displacing the Conflict: Environmental Destruction in Bangladesh and Ethnic Conflict in India," *Journal of Peace Research*, Vol. 33, No. 2, 1996, pp. 189–204.

Swanson, Timothy, Caroline Doble, and Nathalie Olsen, *Freshwater Ecosystem Management and Economic Security*, discussion paper for the June 1999 World Conservation Union (IUCN) Workshop of the World Water Council, November 17, 1999. Online at http://biodiversityeconomics.org/pdf/topics-10-01.pdf as of May 24, 2005.

Thackur, Sunita, "Sardar Sarovar Dam: At What Price Progress?" *BBC News: Special Report: 1998: Water Week*, March 23, 1998. Online at http://news.bbc.co.uk/1/hi/special_report/1998/water_week/65961.stm as of April 8, 2005.

Turton, Anthony, *Water Scarcity and Social Adaptive Capacity: Towards an Understanding of the Social Dynamics of Managing Water Demand Management in Developing Countries*, MEWREW Occasional Paper No. 9, London: University of London School of Oriental and African Studies, 1999.

Turton, Anthony R. and Jeroen F. Warner, "Exploring the Population/Water Resources Nexus in the Developing World," in Woodrow Wilson International Center for Scholars Environmental Change and Security Project (ECSP), *Finding the Source: The Linkages Between Population and Water*, New York: ECSP, 2002. Online at http://wwics.si.edu/topics/pubs/popwawa4.pdf as of April 5, 2005.

UN. See United Nations.

UNCHS. See UN Centre for Human Settlements.

UN CSD. See UN Commission on Sustainable Development.

UNEP. See UN Environment Programme.

UNFPA. See UN Population Fund.

UN, *World Economic and Social Survey 1996*, New York: United Nations, 1996.

———, "New York City Watershed Whole Farm Programme," *United Nations Department of Economic and Social Affairs Division for Sustainable Development*, 1998. Online at http://www.un.org/esa/sustdev/mgroups/success/nyc_wsfp.htm as of April 5, 2005.

———, *World Population Prospects: The 2000 Revision*, Population Studies No. 205, New York: United Nations, 2001.

———, *World Population Monitoring 2001: Population, Environment and Development*, Population Studies No. 203, New York: United Nations, 2002a.

———, *World Urbanization Prospects: The 2001 Revision*, New York: United Nations, 2002b.

———, *World Population Prospects: The 2002 Revision*, New York: United Nations, 2003.

UN, and World Meteorological Organization, *Comprehensive Assessment of the Freshwater Resources of the World*, New York: World Meteorological Organization, 1997.

UN Centre for Human Settlements, *An Urbanizing World: Global Report on Human Settlements*, 1996, New York: Oxford University Press, 1996.

———, *Water Crisis Linked to Poor Governance, Says Toepfer, Briefing to Second World Water Forum*, The Hague, March 2000. Online at http://www.unhabitat.org/hd/hdv6n3/water_pgovn.htm as of April 11, 2005.

UN Commission on Sustainable Development, *Water: A Key Resource for Sustainable Development: Report of the Secretary-General*, New York: United Nations, E/CN.17/2001/PC17, 2001. Online at http://daccess-ods.un.org/access.nsf/Get?OpenAgent&DS=E/CN.17/2001/PC/17&Lang=E as of April 5, 2005.

UN Environment Programme, *GEO-3: Global Environment Outlook 3: Past, Present and Future Perspectives*, New York: United Nations Environment Programme, 2002. Online at http://www.unep.org/Geo/geo3/english/index.htm as of April 5, 2005.

UN Human Settlements Programme, *Human Settlements: Conditions and Trends, Global Urban Observatory and Statistics Unit*, 1999. Online at www.unhabitat.org/habrdd/CONTENTS.html.

UN Population Fund, *Water: A Critical Resource*, New York: United Nations Population Fund, 2002.

UNICEF, "Groundwater: the Invisible and Endangered Resource," 1998.

Vörösmarty, Charles J., Pamela Green, Joseph Salisbury, and Richard B. Lammers, "Global Water Resources: Vulnerability from Climate Change and Population Growth," *Science*, Vol. 289, No. 5477, 2000, pp. 284–288.

Water and Rivers Commission, Government of Western Australia, "Water Facts 10: Groundwater Pollution." Online at http://www.wrc.wa.gov.au/public/waterfacts/10_groundwater_pollution/pollution.html as of May 24, 2005.

Water Science and Technology Board, *Privatization of Water Services in the United States: An Assessment of Issues and Experience*, Washington, D.C.: National Academy Press, 2002. Online at http://www.nap.edu/books/0309074444/html/index.html as of April 8, 2005.

WCD. See World Commission on Dams.

White, Stephen E., "Oglalla Oases: Water Use, Population Redistribution, and Policy Implications in the High Plains of Western Kansas, 1980–1990," *Annals of the Association of American Geographers*, Vol. 84, No. 1, 1994, pp. 29–45.

Whittington, Dale, Donald T. Lauria, and Xinming Mu, "A Study of Water Vending and Willingness to Pay for Water in Onitsha, Nigeria," *World Development*, Vol. 19, Nos. 2–3, 1991, pp. 179–198.

WHO. See World Health Organization.

Winpenny, J. T., "Managing Water Scarcity for Water Security," Food and Agriculture Organization (FAO). Online at http://www.fao.org/ag/agl/aglw/webpub/scarcity.htm as of April 7, 2005.

Wolff, Gary, and Peter H. Gleick, "The Soft Path for Water," in Peter H. Gleick and William C. G. Burns, eds., *The World's Water 2002–2003: The Biennial Report on Freshwater Resources*, Washington, D.C.: Island Press, 2002, pp. 1–32. Online at http://www.pacinst.org/publications/worlds_water/worlds_water_2002_chapter1.pdf as of August 12, 2005.

World Bank, *Resettlement and Development: The Bankwide Review of Projects Involving Involuntary Resettlement 1986–1993*, Washington, D.C.: World Bank, 1994.

———, *World Development Indicators 2002*, Washington, D.C.: World Bank, 2002.

World Commission on Dams, *Dams and Development: A New Framework for Decision-Making: The Report of the World Commission on Dams*, London: Earthscan, 2000. Online at http://www.damsreport.org/report/contents.htm as of April 5, 2005.

World Conservation Union, *Vision for Water and Nature: A World Strategy for Conservation and Sustainable Management of Water Resources in the 21st Century*, 2000. Online at http://www.iucn.org/webfiles/doc/wwrp/publications/vision/visionwaternature.pdf as of April 6, 2005.

World Conservation Union and World Wildlife Federation, *Strategic Approaches to Freshwater Management: Background Paper: The Ecosystem Approach*, 1998. Online at http://www.ramsar.org/key_csd6_iucnwwf_bkgd.htm as of April 23, 2002.

World Health Organization, "Main Findings of the Comparative Study of Social and Biological Effects on Perinatal Mortality," *World Health Statistics*, Vol. 31, No. 3, 1978.

———, *Our Planet, Our Health: Report of the WHO Commission on Health and Environment*, Geneva: World Health Organization, 1992.

World Health Organization and UNICEF Joint Monitoring Programme for Water Supply and Sanitation, *Global Water Supply and Sanitation Assessment 2000 Report*, Geneva: World Health Organization and United Nations Children's Fund, 2000.

World Resources Institute, *World Resources 1996–1997: The Urban Environment*, Washington, D.C.: World Resources Institute, United Nations Environment Programme, United Nations Development Programme, and the World Bank, 1996. Online at http://pubs.wri.org/pubs_pdf.cfm?PubID=2872 as of May 24, 2005.

———, *World Resources 2000–2001: People and Ecosystems: The Fraying Web of Life*, Washington, D.C.: World Resources Institute, 2000. Online at http://pubs.wri.org/pubs_pdf.cfm?PubID=3027 as of August 30, 2005.

World Water Council, "Three Global References Have Been Drafted for Use in the Sector and Regional Consultations," 2000. Online at http://www.worldwatercouncil.org/Vision/ts.shtml as of April 5, 2005.

World Wildlife Federation, *Rivers at Risk: Dams and the Future of Freshwater Ecosystems*, Surrey, UK: World Wildlife Federation, 2004. Online at http://www.panda.org/downloads/freshwater/riversatriskfullreport.pdf as of August 12, 2005

WRI. See World Resources Institute.

WSTB. See Water Science and Technology Board.

WWF. See World Wildlife Federation

WWC. See World Water Council.

Zaba, Basia and Ndalahwa Madulu, "A Drop to Drink? Population and Water Resources: Illustrations from Northern Tanzania," in Victoria Dompka, Alex De Sherbinin and Lars Bromley, eds., *Water and Population Dynamics: Case Studies and Policy Implications: Report of a Workshop: October, 1996, Montreal, Canada*, Washington, D.C.: American Association for the Advancement of Science, 1998.